NEURO TRANSMISSIONS PRESENTS

BRAINS EXPLAINED

HOW THEY WORK & WHY THEY WORK THAT WAY

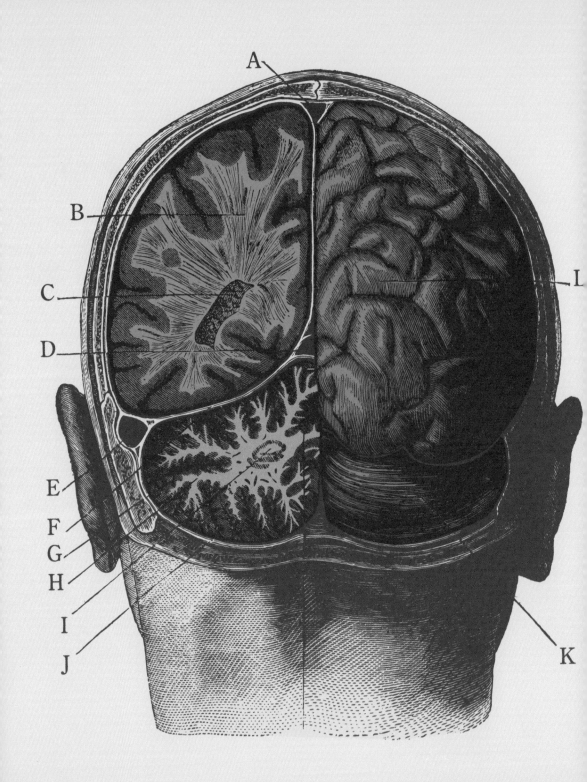

NEURO TRANSMISSIONS PRESENTS

BRAINS EXPLAINED

HOW THEY WORK & WHY THEY WORK THAT WAY

ALISON CALDWELL, PHD
MICAH CALDWELL, LPCC

weldon**owen**

TABLE OF CONTENTS

PART

TWO

MIDBRAIN: YOUR BIOLOGICAL AND SOCIAL BRAIN

PART
THREE

FOREBRAIN: WHERE PSYCHOLOGY AND NEUROSCIENCE ARE GOING

INTRODUCTION

Hey there, Brainiacs! Welcome to Brains Explained

In some ways it seems inevitable that we would have ended up here, making YouTube videos about the brain. Micah has been experimenting with filmmaking since he was in high school, making silly music videos and producing short films with friends over a weekend. And Alie's always been a writer, composing poems as a teen and finishing a whole novel during NaNoWriMo. But when we started out on our career paths as a clinical therapist and a neuroscientist, we never imagined that our two hobbies would collide so beautifully.

In the fall of 2013, when Alie had just started her PhD program, she roped Micah into helping her and her classmates produce a parody music video about the trials and tribulations of graduate school. From that, a friend pointed us to an educational video contest and urged us to apply. We had so much fun making these dang videos that we started to think, "Hey, maybe we're on to something here."

At the time, neuroscience content was hard to come by—and inaccessible. Who wants to watch a boring, hour-long lecture with bad audio? We figured the brain deserved better treatment. And so, in late 2015, *Neuro Transmissions* officially launched on YouTube,

starting with a series of "intro to neuro-science" videos that explained the most basic concepts about the brain, like how a neuron sends a signal and how our senses work. From those humble beginnings, we've expanded to tackle all kinds of ideas and topics in neuroscience and psychology, from the practical (like how to stop procrastinating) to the fantastical (like whether Luke Skywalker's hand could be real) .

To our surprise and delight, people actually like this stuff we make! Five years later, we have been blown away by the enthusiasm and dedication we've received from our viewers. We've heard from adults living with neurological conditions who are thrilled to have a resource to share about what's really going on in their brains and students studying for university exams and teachers excited to have something to use in class. We have viewers in India and Switzerland and all around the world sharing the same love for the brain that we have.

What's not to enjoy? Everyone has a brain, and pretty much everyone knows someone whose brain works a little bit differently. It makes sense that people want to know more about this organ that controls so much of our lives. Our brainy audience has continued to grow every day—and now, with this book, maybe our audience will grow a little bit more. *Brains Explained* is a new format for us, and one that we hope will help many new brainiacs have fun while better understanding neuroscience and psychology and how it affects their daily lives.

In the following chapters, you'll learn about the brain from front to back and side to side, from our distant past and on into our brilliant future. We'll talk about how the brain, and the ways we think about it, have evolved over time. We'll discuss dubious medical treatments for mental illness and futuristic new therapies to boost your brainpower. You'll learn how to figure out if your cat really loves you, why you need a therapist, and whether or not those "brain games" are really worth it.

If you're already a subscriber, thank you for coming along on this ride. Our work—and this book—would not be possible without you. We're so grateful that you want to hear what we have to say, and we hope you're excited to dig a little bit deeper into your brain. And if you're new here, welcome. Buckle up, because you're about to learn all about those billions of big, beautiful brain cells, and how they all come together to make . . . *you*!

– *Alie & Micah Caldwell*

PART
ONE

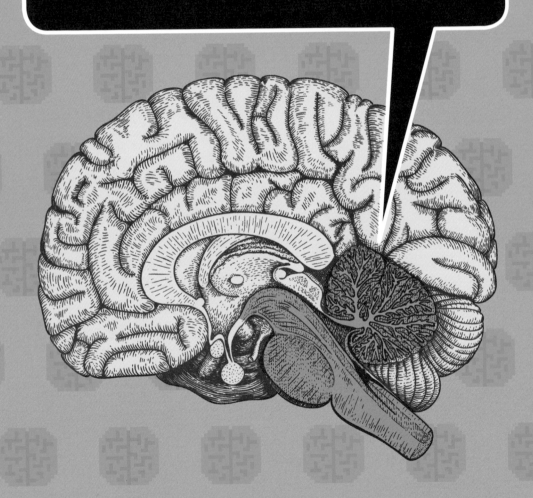

The name "hindbrain" is a bit tongue in cheek; in neuroscience, the hindbrain is used to describe a group of structures located on the brain stem, including the medulla, pons, and cerebellum. But don't worry, we're not going to spend a whole bunch of time explaining brain anatomy to you . . . as exciting as Alie might think that would be.

This first section, Hindbrain, is all about hindsight—we're looking backward in time, catching you up to speed on how our brains evolved to think about themselves, and all the weird and wonderful (and sometimes awful) things neuroscientists and psychologists of yesteryear did to try and understand them.

Along the way, we'll debunk some brain myths and talk about some important historical moments . . . including the truth about your so-called "lizard brain," which philosophers thought the brain had something to do with sperm, and why lobotomies were sooo popular in the early 1900s. And you'll get to meet some real characters, like the unkillable Phineas Gage, Harlow and his monkeys, the electrifying Galvani and his frogs, and of course, Pavlov and his dogs.

But to begin at the beginning—at one point in time, billions of years ago, brains didn't exist. And then, eventually, some little bitty single-celled organisms, called eukaryotes, evolved to use electricity to send signals. And with that little spark, evolution was off to the races.

Somehow, over the course of two billion years, we got to where we are today: sitting in our sweatpants at our desks, cramming Cheetos into our mouths and desperately trying to finish this book on time. So what happened during those two billion years? And how has the work of neuroscientists and psychologists throughout history helped us understand the brain—or, as the case may be, how have they misled us?

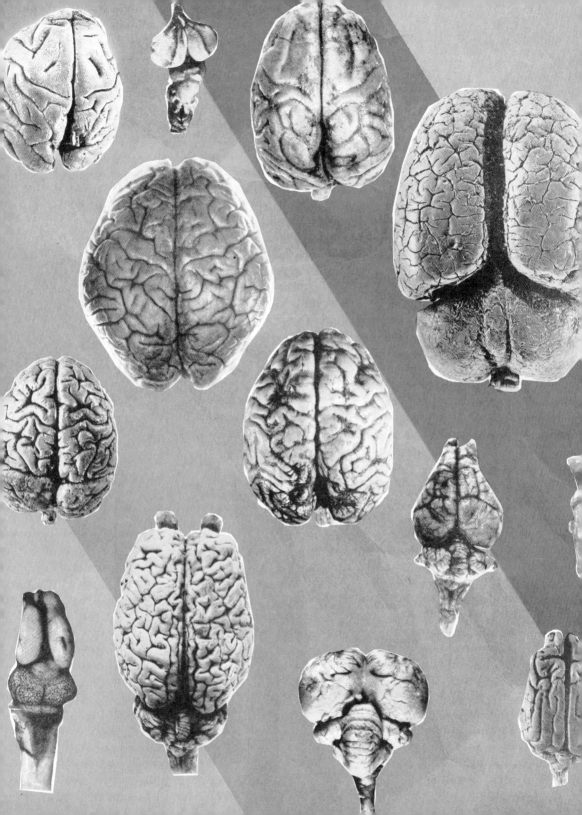

Chapter 1

WHEN WE SAY LIZARD BRAIN . . .

. . . we don't *literally* mean lizard brain. Even with all our modern, newfangled "science," our brain's evolutionary history is something of a mystery. Understanding the brain is no small feat, but part of figuring out how it works involves figuring out where exactly it came from (spoiler: not aliens.)

So let's start our journey through the brain at the very beginning. It all started long ago, in an ocean far, far away . . .

Brains are hard to study. They're fragile, and mushy, and start to fall apart pretty much the minute we die. And they don't really fossilize well, either. So how on earth *did* 4.5 billion years take life all the way from single-celled organisms to us sitting in our sweatpants trying to write this book while the cats walk across our keyboards? And how do we know?

Centuries of scientists have endeavored to answer this question because understanding how our brains evolved might hold the key to

understanding how—and if—our big, wrinkly human brains are unique among the species. At something other than meme production, anyway.

Without many fossils, it's hard to figure out how any of the thousands of different brains on the planet evolved over time. One way we can start to understand brain evolution, though, is by comparing the brains of various species that exist today: How similar our brains look to our closest animal relatives, compared to a lion's, or a tiger's, or a bear's . . . oh my!

HOW BRAINS EVOLVED

Brains have been around almost as long as life has—there's evidence that even some early single-celled organisms used chemical and electrical signaling processes similar to what our nervous systems use today. So how did we go from single cells to Single Ladies? (If you liked it, then you should've put a brain in it . . . we'll see ourselves out).

600 MILLION YEARS AGO
That Magic Touch. The first "nervous system," called a nerve net, allows creatures like jellyfish and sponges to respond to touch.

FUN FACT: Some species of invertebrates (like sea squirts) actually *lose* their brain during development, with the brain dissolving once the larvae find a permanent location.

2 BILLION YEARS AGO
It's Electric! Single celled eukaryotic organisms evolve the ability to generate an electrical charge.

Note: A billion years is a really, really long time. Of course you know that, but we mean, like, a really long time. As in, if we actually represented this timeline in proportion, you'd have to have a several hundred page foldout of . . . nothing between these two entries. And that would have used up a lot of trees, so we're hoping you'll forgive the inaccurate scale!

550 MILLION YEARS AGO
The First Brains (Kinda). The first brain probably evolved in a worm, but we're not sure, because we've never found a fossil—remember that bit about turning into mush?

550 MILLION YEARS AGO
Something's Fishy Here. Around the same time, fish (sort of), along with their brains (sort of), arrived on the scene, along with the spinal cord. Based on studying modern-day species, we think that this is around when the brain started splitting up into different regions—the forebrain, midbrain, and hindbrain. All species of vertebrates have brains that grow from those three early structures.

400 MILLION YEARS AGO
Super Mutants. Around when amphibians first took our brains on land, one of our ancestors accidentally duplicated its entire genome, which was a huge leg up for evolution—it gave our genetic code a lot of space to try new things and see what worked. This gave the brain a chance to try out new chemical signals, which made it possible for animals to do more complex behaviors.

200 THOUSAND YEARS AGO
Actually Human! Okay, for real, it's really us, finally! The first anatomically modern humans, that is. If you dropped a human being born 200,000 years ago into today, they would be just as capable and smart as any current human—if they weren't left completely terrified by the modern world.

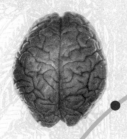

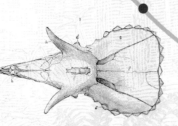

230 MILLION YEARS AGO
Clever Girls. Dinosaur brains were probably relatively similar to modern day animals—in that their brains would vary a lot based on the species and their behaviors.

Stegosaurus' brains weren't really the size of a walnut—our apologies to your inner ten-year-old—but they were proportionally very small when compared to their size. Imagine a dog's brain—in an animal the length of a school bus.

On the flip side, velociraptors were probably not as smart as Jurassic Park would have you believe—many carnivore dinos had brains that were "average" compared to modern-day lizards.

225 MILLION YEARS AGO
Not So Special Now, Are We? The first mammals arrived, evolving from amphibians and existing alongside dinosaurs, and TBH there's not a whole lot about their brains that are unique just because they're mammals—the uniqueness is more evident when we start looking at specific mammalian lineages.

6 MILLION YEARS AGO
Almost Human. Okay, we're almost to us, we swear. Brains started getting bigger and better as the first hominids, our ape-like ancestors evolved, and bigger brains allowed for skills like tool use, which made it easier for us to get food, which meant our brains could go bigger . . . it was like a vicious cycle, but, you know, good.

All modern animal species have been evolving just as long as we have—we're just more closely related to some of them than others. How closely? Glad you asked, because that allows us to use the word "cladogram," the scientific term for this very sort of diagram that shows how various species relate to each other. This particular cladogram shows you the relationships between different species, and how our brains stack up compared to our cousins. Each split show a place where the lineage diverged, bringing us to the species that exist today.

ONE BIG BRAINY FAMILY

Yes, as a **human**, your brain is *very* special. Well, so we like to think. But when you get right down to it, our brains are pretty much just like every other animal's, but maybe a bit more wrinkled. And maybe a bit more worried about finding a prom date.

Rodents rely heavily on their sense of smell—so much so that they have a specialized lobe on the front of their brain, called the olfactory lobe, whose only job is to process smell information.

Cats may think they're better than you. You may think cats have tiny, stupid brains. (You monster.) But they're more like humans than you might imagine. They have object permanence, which human infants don't even have, and complicated dreams. In fact, scientists think they could be smarter than we know—they're just not very cooperative in the lab and don't really care about our weird experiments.

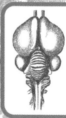

Reptile brains aren't really primitive—reptiles have been evolving just as long as we have. That's right, according to two billion years of evolution, Elon Musk is completely equal to a gecko. Any gecko. Every single gecko.

Scientists think that **frog** brains are more similar to fish brains than they are to reptile brains. And evolution doesn't think these guys need to be brainy—some species, like salamanders, actually seem to have had their brains shrink over time.

Turns out that being called "**bird** brain" isn't such an insult—birds may have small brains, but their neurons are packed much more densely than those of humans, so their brains are more like ours than you'd think!

Bony fish, which are basically little minnows and not nearly as cool-looking as they sound (although you can buy genetically engineered zebrafish that glow different colors), can actually regenerate some parts of their brain after injury, making them super interesting to study. Imagine the sci-fi-esque medical advances! Imagine!

Despite having small, strangely shaped brains, some species of **shark** are actually pretty smart—and some even seem to be able to use their metabolism to "heat up" their brains above ambient temperature, to make their vision and hunting skills sharper.

THOSE BRAINY OCTOPUSES

Sometimes evolution gets wacky. Like the time the ancestors of octopuses went, "You know what's better than one brain? Nine!"

So Many Brains

We're so used to thinking of our own brain as *the* brain, it's pretty easy to forget that vertebrates aren't the only cerebrally-endowed creatures around. *Most* animals have brains, and they can be pretty similar to our own—or totally, wildly different. Like octopuses, who do in fact have nine brains. Sort of. The octopus's brain is actually shaped like a donut—it's wrapped around their esophagus, and they have to be careful about what they eat, because if something is too big, it'll bump into their brain on the way to the stomach. Imagine taking too big of a bite of cheeseburger and giving yourself a brain injury!

But octopuses *also* have smaller "brains" (actually called ganglia) in each of their eight arms, giving them fine-tuned control over their tentacles and deft, sensitive suckers.

This "network" of brains is thought to be why octopuses are so smart. They can solve puzzles, break out of their tanks, and open jars; for a lowly invertebrate, they're actually pretty human-like.

Size Doesn't (Always) Matter

Octopus brains are part of a growing body of evidence that challenges our assumption that bigger brains are always better (smarter) brains. Octopuses only have about 500 million neurons, compared to humans, who have nearly 100 billion—but researchers think that because of their networked brains, octopus tentacles can almost "think" for themselves, and that might be part of why octopuses seem so smart.

But who cares, right? It's not like we're going to be transplanting octopus brains into our own heads any time soon. (Well, *we're* not, anyway.)

Ongoing Studies

Octopus brains—and the brains of other invertebrates—are actually surprisingly useful for understanding our own brains, and figuring out how (and if) we're unique. All brains use similar underlying principles—they're information hubs that send electrical signals across the body to help an organism react and move, and use a variety of chemical messengers to do their bidding.

So how are our brains clever enough to cut open the skull and take a look at that smart, squishy organ to better understand it, while other species seem to have a hard time with abstract thoughts and ideas?

Studying other animals, even the ones whose brains are super different from ours, helps us understand how they behave and think—and knowing the similarities and differences between us and them gets us a little closer to understanding what it actually means to be human.

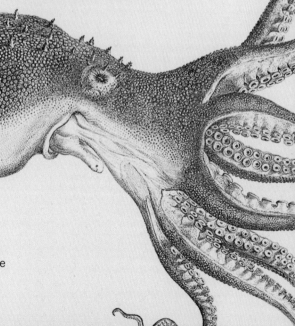

THE LIZARD BRAIN

There used to be this theory that evolution treated our brain like Legos—it just kept piling new bits and pieces on top of the old ones as we evolved from fish into lizards into mammals into humans.

Back in the 1960s, this neuroscientist named Paul MacLean had the idea to classify our different brain regions based on their "base" functions.

The Misconception

Folks used to think that our "lizard brain" controlled our primitive instincts, while our "homo sapiens" brain was responsible for all our smarts.

According to MacLean, the protoreptilain brain—or the lizard brain—was the oldest structure, consisting of the brainstem and cerebellum. It was thought to be responsible for the different functions that keep us alive, like breathing and sleeping, and was considered to be very simple and unchanging. It responds only instinctively, animal-like, with no thought or emotion.

The paleomammalian brain included the limbic system, a set of connected brain structures including the amygdala, hippocampus, and hypothalamus, which are important for motivation and emotion—like the motivation it takes to keep all your kids alive. This was the "emotional" brain, reacting to the world with raw, unfiltered feeling, and telling us if things feel good or bad.

Finally, the neocortex was bundled into the concept of the neomammalian brain, as this outermost brain layer is found only in mammals, and is critical for our highest level cognitive abilities—decision making, planning, talking, using tools, and so on. This is our logical, rational brain, overseeing the other two.

Why Did We Think That?

The theory came from comparative neuroanatomy, in which scientists try to figure out when and how different parts evolved by comparing the brains of different species. But it turns out that our understanding of the evolutionary relationships between species wasn't quite up to snuff in the 1960s.

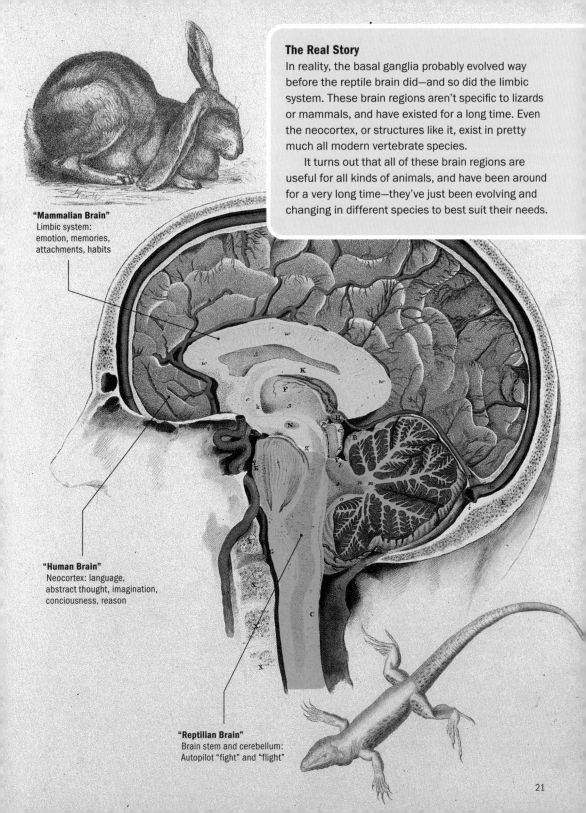

"Mammalian Brain"
Limbic system:
emotion, memories,
attachments, habits

The Real Story

In reality, the basal ganglia probably evolved way before the reptile brain did—and so did the limbic system. These brain regions aren't specific to lizards or mammals, and have existed for a long time. Even the neocortex, or structures like it, exist in pretty much all modern vertebrate species.

It turns out that all of these brain regions are useful for all kinds of animals, and have been around for a very long time—they've just been evolving and changing in different species to best suit their needs.

"Human Brain"
Neocortex: language,
abstract thought, imagination,
conciousness, reason

"Reptilian Brain"
Brain stem and cerebellum:
Autopilot "fight" and "flight"

OTHER BRAINY CRITTERS

While invertebrates can be more clever than you might expect, it's still usually true that the bigger the brain, the smarter the critter. Which critters are the smartest? Or, at least . . . the brainiest? First, let's take a little look into the past to see where some of this braininess may have begun.

Dinosaurs

While some dinosaurs did have pretty small brains (looking at you, stegosaurus), members of groups like the *troodontidae* family had brains that were nearly as big as the brains of modern birds. And no, none of them had two brains. That was just a super awkward misunderstanding based on a super weird bit of stegosaurus anatomy.

Neanderthals

Neanderthal brains were actually bigger than human ones. We like to think of Neanderthals as our more primitive cousins, but Neanderthals were quite smart—probably at least as smart as us. There's evidence that they created tools and art, and may even have buried their dead.

Mammoths

In 2010, a 40,000-year-old wooly mammoth carcass was discovered frozen in the Siberian permafrost, so well preserved that its brain was almost completely intact. Scientists studying the brain said that the mammoth's brain was basically the same as a modern African elephant, and that they were probably just as smart.

And now, let's look at the brains you can still find out there in the world today.

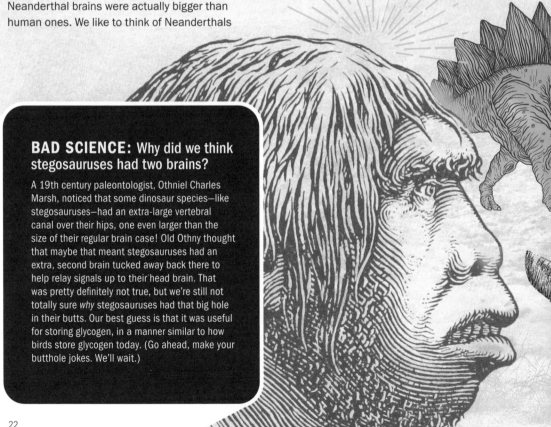

BAD SCIENCE: Why did we think stegosauruses had two brains?

A 19th century paleontologist, Othniel Charles Marsh, noticed that some dinosaur species—like stegosauruses—had an extra-large vertebral canal over their hips, one even larger than the size of their regular brain case! Old Othny thought that maybe that meant stegosauruses had an extra, second brain tucked away back there to help relay signals up to their head brain. That was pretty definitely not true, but we're still not totally sure *why* stegosauruses had that big hole in their butts. Our best guess is that it was useful for storing glycogen, in a manner similar to how birds store glycogen today. (Go ahead, make your butthole jokes. We'll wait.)

Sperm Whale

One way that scientists try to estimate animal intelligence is by looking at an animal's brain-to-body size ratio. Humans are pretty far up on that list, as are elephants, chimpanzees, and pigs—but the actual winner in this category is the sperm whale, which has a brain about six times as big as a human's. Their brains have extra-large sound processing regions—all the better to echolocate with!

Bonobos

Sure, chimpanzees are our closest living relatives—but it turns out we share just about as much of our DNA with bonobos, our smaller, gentler cousins. Both species are pretty smart, with complex social lives and the ability to use tools. But bonobos' brains seem to be better than chimps' at handling some kinds of social interactions, like controlling aggressive impulses and understanding when someone else is upset. Which is probably why bonobos are less likely to kill you.

Crows

Crows have pretty big brains, for birds. And even though birds never evolved a neocortex—they split off from mammals looong before that—they did evolve a structure called the nidopallium, which is basically the same thing. Their superior smarts mean that some species of crows can perform impressive cognitive feats, like remembering human faces and using tools.

THEORY OF MIND

How do you know you're you? And how do you know I'm me?

Your ability to understand that you have your own beliefs and desires—and that I have mine—is called "theory of mind."

Theory of mind is a critically important part of what it means to be human, because without it, we'd never give a fart what anyone else thought or felt, and society would fall apart pretty quickly.

Mirror, Mirror, on the Wall

We think that some other species have theory of mind, at least to some extent. For example, most of the great apes pass the "mirror test"—basically the animal version of drawing a moustache on your friend's face while they're sleeping. If an animal sees itself in a mirror, notices the moustache, and then touches its own face, it shows that the animal understands that it's seeing its own reflection. Scientists interpret this as meaning that they have a sense of their "self."

Other animals that pass the mirror test: dolphins and killer whales, elephants, Eurasian magpies, and even a fish called the cleaner wrasse. Other kinds of testing show that other species, including dogs, pigs, ravens, and goats, seem to be able tell the difference between themselves and other animals.

I, Human

If theory of mind is part of what it means to be human, what does it mean if other animals have it, too? Well, even if some of our tests seem to indicate that other animals have theory of mind, it's really hard to know what exactly is going on in the brains of those other creatures—at least until we figure out how to build a bark-translator machine. Some scientists even argue that language and theory of mind go hand in hand, so without language, theory of mind may not really exist. So we can't really prove that they understand that I'm me and you're you and they're them.

But as we better understand different animals and the kinds of things they're capable of understanding and feeling, it's worth considering how we treat and study these critters—and how we should be changing our own lifestyles to protect theirs.

SO, WHAT ABOUT OUR ANCESTORS, ANYWAY?

As humans, we like to think we're special, because we're the only animals who do fancy things like wear clothes and drive cars and speak languages. But that almost wasn't true. There were a whole bunch of human-like species hanging out for a few million years. We just happen to be the only ones to have survived for this long.

Because all of our closest cousins are long dead, we can only make guesses about their brains by studying the shapes of skulls and the imprints of blood vessels left inside the fossilized skulls that we have found.

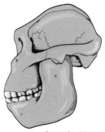
Australopithecus

Around 5 million years ago, ancient humans evolved the ability to walk upright and started figuring out how to use simple tools—but their brains weren't particularly big yet.

Then, about 2 million years ago, the brawnier *Homo erectus* appeared on the scene with a much bigger brain to match. *H. erectus* is thought to be the first species to make their own stone hand axes, control fire, and maybe even create art.

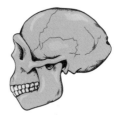
Homo erectus

Our extinct relatives, Neanderthals, showed up around half a million years ago, with the biggest brains of any hominin before or since. By all appearances, they were pretty much just like us, if a bit more hairy. They could make clothing, build boats, and even care for each other's injuries. Oh and uh . . . the reason some of us have a little bit of Neanderthal DNA? Yeah, it's not because we're descended from Neanderthals; it's because ancient *Homo sapiens* and Neanderthals totally boned.

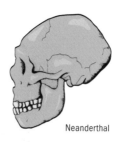
Neanderthal

Unfortunately, we'll never get to find out what our cousins were actually like, because they went extinct about 40,000 years ago, and . . . it might have been our fault. While a small population and fragile genetics probably contributed to their demise, there's also evidence that early humans may have butchered and even eaten Neanderthals. How's that for brotherly love?

Homo sapiens

OUR BIG, BAD BRAINS

Our brains are pretty ridiculously huge, compared to many other species. Combined with our narrow pelvis (thanks for that, bipedalism), it means that birth is pretty rough on humans. So evolution decided to help us out by allowing us to birth human babies while they're still super soft and squishy. This means that our brains still have a ton of growing to do after birth, which is why we're so helpless and hilariously stupid when we're babies.

Chapter 2

ARISTOTLE SAY WHAT?

Brains: They're pretty great, right? From ancient times onward, we've known that the brain was taking up a fair amount of real estate in the body. But we didn't always know what the darn thing was actually *used* for.

The brain is a cold, moist organ that produces sperm . . . at least, that's what it was if you listened to Galen, one of the ancient world's most venerated scholars. Yep, the study of the brain took some, uh . . . pretty unexpected turns through history. But ya know what? Those kinds of bold statements and terrible hypotheses from the past are part of what propelled us to get to where we are today in our scope of knowledge. Humans are not born with any real inherent understanding of the mind and, to be honest, it's pretty incredible that any brain can be powerful enough to study itself. Think about it—that would be like a robot becoming conscious and aware of how it was built, and then studying its own CPU.

(We do in fact think about that idea as well, as you'll see in chapter 18).

That's essentially what happened—throughout the ages and around the globe, many human brains have studied other human brains (as well as their own). Some of those brains came up with harebrained ideas, like the ol' "sperm comes out of your brain" theory. But other brains came up with much better big-brain ideas! Some of them were pretty revolutionary, like drilling a hole in your skull to treat a brain injury . . . or talking about your problems with a trained professional. Along the way, we got rid of the bad brain stuff and built on the good brain stuff and here we are!

A BRIEF BRAINY HISTORY

The study of the brain began a loooooong time ago. We couldn't keep people from messing with that smart, squishy stuff. But as they say, it takes a village—or in this case, an entire world—to finally figure out as much as we have. (Read on; there's still a lot we don't know!) From the beginning of written history to medieval times, lots of cultures around the world have formed about the brain, though sometimes their practices (and their conclusions) were . . . questionable.

Renaissance Italy
For the first time in centuries, human dissection was allowed and some artists and scientists were not afraid to get their hands dirty by cracking open the skull to see what was hiding under there. In between painting the Mona Lisa and inventing helicopters, Leonardo de Vinci also studied the brain and made some important observations.

Ancient Incan Empire
Holey moly! Hundreds of Incan skulls have been found with holes drilled in them . . . and we can see in the way the bones healed that those people survived.

Ancient Greece

Socrates bragged about having hallucinations that would tell him to do things. He called this inner voice his "daemon." Cute or nah?

Ancient China

Near the beginning of the 3rd century, Han Dynasty leader Cao Cao is said to have called upon a prominent doctor named Hua Tuo to treat his terrible headaches. Hua Tuo, known today as the father of Chinese surgery, was already famous for his healing powers. The story goes that he suggested brain surgery (or probably something more like trepanning) to Cao Cao, who took it as a death threat and had him executed. Headaches make everyone so cranky, right?

Baghdad's Golden Age

In the 9th century CE, the influential Persian Muhammad ibn Zakariya al-Razi, known as Rhazes, was one of the first scholars to describe mental illness and an early form of psychotherapy, and became director of one of the world's first psychiatric wards.

Ancient India

Buddhist and Hindu beliefs promoted the balance of mind and body with meditation and yoga. Recent studies show there may be something to the "Sanskrit effect," a cognitive benefit of memorizing all those mantras.

Ancient Egypt

As part of the mummification process, ancient Egyptians took out all your insides except the heart. Why? They viewed the heart as the seat of all our thoughts and emotions. Whoops!

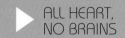
Ancient Egyptians sure loved their mummies. Recent evidence shows that they started doing artificial mummification as far back as 4500 BCE. They believed that preserving your body here on earth meant a sexy, healthy body in the afterlife. But it's hard to get the perfect leathery skin with all of those decaying organs sitting in your body, so Egyptians started taking out all our insides—liver, kidneys, intestines, and yep, you guessed it: the brain.

Talk About A Brain Drain

The brain removal process was simple, but pretty gnarly. Don't try this at home! Essentially, they stuck a long stick or iron hook up the nose, and punched a hole through the bone separating the nasal cavity from the brain. Delightful! Then they swished it around a few times and pulled out a bunch of the insides. After getting most of the chunks out, they tipped over the body and let the rest drain out. Sounds like a really thorough neti pot rinse.

Egyptian Cranial Crochets!
(used for brain removal)

Be Gone, Devil!

Early Egyptian medical papyrus scrolls were surprisingly sophisticated. The Edwin Smith papyrus is the first place the word "brain" is ever used, and its author compared the surface of the brain to "molten copper." The Eber papyrus, meanwhile, described how to treat depression,

though the Egyptians viewed it as an ailment of the heart, not the brain. That may explain why they threw the brain away during mummification. But for all their sophistication, they also had some dubious remedies—incantations, spells, and witches' brews. Sorcerers were on par with physicians, and thus religion, medicine, and magic were intertwined and indistinguishable.

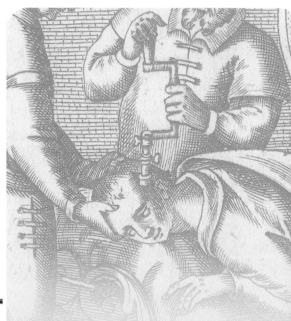

WEIRD SCIENCE: TREPANATION

In 1867, the American ambassador to Peru encountered an ancient Incan skull with a rectangle-shaped hole in it. What's wild is that it was cut out on a living person . . . and that the person survived! Drilling a hole in your head sounds like a bad idea. And yet, we have hundreds of skulls that were "trepanned" all over the world, dating as far back as 6500 BCE and as recently as 2000 CE (illegally). Sometimes it was done for ritualistic purposes or as a cure for mental illnesses or after traumatic brain injuries or for drainage—which sounds gross. It is a legit medical procedure for relieving pressure on the brain after injury. Today, we still cut holes into people's skulls for surgery or treating head trauma, but we tend to put the bone back afterward.

The ancient Greeks were obsessed with finding the location of the human soul or, as they called it, the *psyche*. But they couldn't make up their goshdarn minds about where it lived, so you had some serious disagreements between physicians and philosophers (not entirely two different job descriptions at the time, just to make it all a little more confusing).

The Physicians

The human body was sacred in ancient Greek culture and, as such, dissections were a big no-no. This meant that the brain's anatomy was largely a mystery, but multiple physicians, including Hippocrates, did correctly identify it as the seat of intelligence. Hippocrates even described several mental disorders, perhaps corresponding to depression and anxiety, but he attributed them to an imbalance of the four "humors," or bodily fluids (all the rage at the time diagnosis-wise): black bile, yellow bile, phlegm, and blood. Even so, this was a big improvement on the prevailing theory of the time: that mental disorders were a punishment from the gods or perhaps a sign of demonic possession.

The Philosophers

Aristotle sought the source of what he called *aísthēsis koine*, or "common sense," though the term meant something pretty different than how we use it today. Rather, he was imagining a literal location in the body that integrates all of our senses and memories, along with rational thought. But even with the evidence coming from his physician friends that the head might be where it's at, Aristotle was like, "It's definitely the heart tho." So what did he make of the brain? Dude thought the brain was used for cooling the blood. Indeed, he hypothesized that this is why we're more rational than other creatures—because we have larger brains to cool down our hot-bloodedness. For a man who prided himself for his superior logic skills, he sure made some wild logical leaps here.

THE HIGHEST LADY IN THE LAND: DRUGS AND PROPHESY

The Oracle of Delphi was believed to be the mouthpiece of the god Apollo. Emperors and peasants alike visited the oracle's vapor-filled temple in search of divine prophecies. Its ruins remain, but scholars thought the vapors were myth, as they saw nothing that would explain the phenomenon. But in 2001, a geologist, a chemist, a toxicologist, and an archaeologist turned up some fascinating evidence (after, we hope, walking into a bar together): The rocks under the temple are made of oily limestone and fractured by two previously undetected faults—a recipe for a temple filled with sweet, sweet ethylene gas (no really, it smells sweet). When inhaled, ethylene induces a trance-like state that affects your speech patterns and body control, while letting you remain lucid. So it turns out she was just reeeeeeally high. Party at the temple!

Galen was the big cheese of natural science in the Roman Empire and, in the spirit of the Romans, he copied a lot of his ideas from his Greek counterpart, Hippocrates.

What He Copied

Like Hippocrates, Galen thought that rational thought dwelled within the brain. But Galen one-upped him and said there was no difference between the mental and physical, which meant psychological issues were not demons or the gods. In fact, Galen made huge advances in psychology by suggesting counseling as a way of treating people, rather than administering dubious elixirs or locking them up. He also loved the four humors, and in a not-so-great move, Galen diverged from Hippocrates and proposed that personality issues were the result of an imbalance in your bodily juices. The solution? Bloodletting, of course! Just drain that blood and soon enough you'll be good as new. It's amazing this practice remained so popular until the 19th century.

What He Invented

Some of Galen's ideas he came up with on his own . . . like the theory that sperm comes from your brain. No really, he thought there was a direct pathway from your head to your junk. When someone gets hot and bothered, Galen believed, your brain excretes semen into the spinal cord, where it mixes with the fatty residue from your other body juices as you warm up. This foamy substance then flows down your spinal cord and into your loins. Sure, why not? And although he believed the brain was the seat of intelligence, he believed we actually have three souls: the rational soul in our brain, the spiritual soul in our hearts, and the appetitive soul in our liver. (I dunno about you, but when I'm hungry, my appetitive soul definitely possesses me.)

In ancient times, they didn't use the term "consciousness." Instead, they referred to it as the "soul," "psyche," "essence," or "life force," among other terms. Turns out, there were a lot of wild ideas of where that hard-to-define quality could be found.

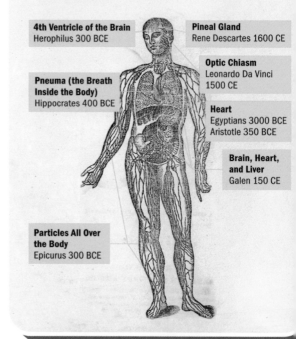

4th Ventricle of the Brain
Herophilus 300 BCE

Pineal Gland
Rene Descartes 1600 CE

Optic Chiasm
Leonardo Da Vinci 1500 CE

Pneuma (the Breath Inside the Body)
Hippocrates 400 BCE

Heart
Egyptians 3000 BCE
Aristotle 350 BCE

Brain, Heart, and Liver
Galen 150 CE

Particles All Over the Body
Epicurus 300 BCE

Chakra It Up to Fate

Galen's idea that the body's humors change your personality was odd, but honestly was nothing new. About 500 years earlier, Hindu healers in India had already come up with a similar theory that your physical health, mental health, and personality were affected by an imbalance of three body substances. Medieval Indian gurus evolved that belief into the concept of chakras, which are nodes of psychic energy. That's right, we're getting mystical! When the flow of energy from these chakras becomes blocked, the body, mind, and emotions get all messed up—supposedly. But techniques like breathing exercises, yoga, and meditation allowed you to control them and obtain the ultimate mind-body connection. The idea of mind-body connectedness has been widely incorporated in therapy and medicine. Just look out for the smug American yogis who have adopted chakras as a pseudo-religion.

The Natural and the Supernatural

Ancient China's approach to the mind was holistic. They believed that balance and harmony between the body, mind, and nature were necessary for good health. So logically, mental illnesses were thought to be caused by physical and environmental abnormalities like an inflamed liver or a heatwave. Later, due to religious influences, Chinese medicine incorporated supernatural elements and got a bit obsessed with demonic possessions. They believed that when a person experienced trauma or became emotionally overwhelmed, that made them more vulnerable to evil spirits literally entering their body and causing blockage. To treat these issues, people often turned to herbal medicines, acupuncture, and even the excessive expression of emotions to just get it all out there. Sort of like today's rage rooms—got some excess anger you need to vent? Book a smashing-and-screaming session now!

BAGHDAD'S GOLDEN AGE

While Medieval Europe largely went through a period lacking in much advancement, the Islamic caliphates were booming with new discoveries. Physicians like Avicenna began performing surgeries on head injuries and skull fractures to learn more about how the brain worked.

A Man Ahead of His Time

This Avicenna guy was something special, too. He wrote the first medical encyclopedia, which was fourteen volumes long. Talk about commitment! In it, he identified a wide variety of neuropsychiatric conditions—hallucinations, dementia, insomnia, and epilepsy, to name a few. He also took a medical approach to mental illness and treated it as a physiological and emotional issue instead of evil spirits or possession. In fact, Avicenna was probably one of the most influential people in medicine you've never heard of.

Razi Dazzle 'Em

The Islamic Empire was a hotbed of amazingly smart and scientific physicians. But one of those physicians, Abu Bakr Muhammad ibn Zakariya al-Razi—call him al-Razi for short—was probably the most influential in the field of psychology. This intellectual badass wrote two hundred and thirty-seven books on medical observations he made, including psychological issues such as depression, anxiety, schizophrenia, and mania. He essentially wrote the handbook for identifying common symptoms and treatments for mental health problems. Al-Razi thought that mental illnesses should be treated like medical issues, which led him to create the first psychiatric ward ever, in the hospital he ran. These later proliferated and spread throughout the region, leading to relatively advanced perspectives on mental health. Forget the image in your head of an "insane asylum": The mentally ill were treated with respect, and received medicines, music therapy, and job training. Al-Razi even developed a primitive form of psychotherapy where he spoke to patients on a personal level in a gentle, hopeful way. It's almost like treating people with dignity can improve outcomes! Who knew?!

MEDIEVAL EUROPE

If you lived in Europe during the Middle Ages, then having a mental illness was . . . challenging. Early Christian theology relied on supernatural explanations, and there was a very thin line between "holy" and "madness." Monks engaged in "holy anorexia" and religious figures like St. Francis and St. Joan of Arc reported experiencing spiritual visions. Not to downplay the importance of religious figures, but these saints may have had religious hallucinations, which can occur as a symptom of schizophrenia. Due to their religious nature, they were deemed divinely inspired. However, others were not so lucky.

She Turned Me into a Newt!

Most often, psychiatric issues were seen as a spiritual struggle against evil. If you lost the battle, then you quite literally "got the devil in ya" and succumbed to madness. People who were possessed by the devil were considered witches, and there's only one thing you can do with a witch. Burn 'em! It's estimated that around sixty thousand witches were put to death during this time. But this wasn't the only medieval "cure." To drive out the devil, suggested treatments included prayer, whippings, and exorcisms. If that didn't work, then these poor souls were tortured or, as mentioned before, burned alive.

YOUR BRAIN ON... ALCOHOL

What It Is: Alcohol

What Type of Drug It Is: Depressant

What It Does: Anything from jovial bonhomie to organ failure and death.

How It Does This: It's a myth that Medieval people feared drinking water. However, beer and wine were so prevalent with meals back then that alcohol was practically its own food group. But what's all that alcohol doing to your noggin'? After you slam a Jager bomb, the alcohol binds to GABA receptors. GABA is an inhibitory neurotransmitter that prevents neurons from firing. GABA receptors are concentrated in the prefrontal cortex, the hippocampus, and the cerebellum, so high levels of alcohol depresses these areas most, resulting in common side effects like poor decision making, difficulty remembering the night, and bad balance. Alcohol also increases levels of norepinephrine, the neurotransmitter responsible for arousal, which can increase impulsivity and make you seek out pleasure. No wonder you want Taco Bell at 2 a.m.!

What the Risks Are: Aside from the risk of alcohol poisoning, booze can be highly addictive. Longtime users may experience severe withdrawal—including potentially fatal problems with circulation and breathing.

A WHOLE LOTTA NOPE: EXORCISM

Exorcisms get a bad rap (at least when it comes to mental illness). Being possessed by another being is a belief as old as time. In many religions and cultures, it was not always seen as a bad thing and was often used to describe behaviors or thoughts that seemed uncharacteristic or odd. Grandparents start babbling nonsense; innocent children start dressing differently, and disobeying their parents; chaste spouses are discovered having affairs with the neighbor. Unsurprisingly, mental illnesses like depression or schizophrenia were also seen as a product of possession by evil spirits.

The Power of Christ Compels You!

Christianity, however, took a staunch view of possession as an exclusively bad thing. If a person was possessed, that meant they

were taken over by a demon that must be cast out in order to save the person. So, to drive these spirits out, Catholicism introduced the rite of exorcism in 1614. The rite itself is fairly simple—a priest prays over the afflicted individual, invokes the name of God, and commands the demon to leave. Although often harmless, sometimes it involved tying the individual down if there was a fear of violence to self or others.

The Exorcist Reloaded

The rite of exorcism had all but disappeared for over two hundred years until *The Exorcist* came to the big screen in 1973. The film was based on a real-life exorcism of a young boy in the 1950s. Suddenly, everybody wanted an exorcism for their supposed demonic affliction and the requests for exorcisms have remained high ever since. It was such an issue that, in 1999, the Vatican had to issue new guidance for when to conduct exorcisms. Today, officially sanctioned exorcisms are extremely rare, but there is a blossoming black market for under-the-radar exorcisms (as well as DIY exorcisms which can sadly involve dangerous and deadly practices).

So . . . Why Defend Exorcisms?

This is an unpopular opinion, but the rite of exorcism arrived at a time when our understanding of mental illness was primitive. Treatment for mental illnesses were often ineffective, harmful, or non-existent. Viewing depression or anxiety as an affliction by demons actually destigmatized mental illness by attributing it to external forces rather than personal failings. There is no scientific evidence for the effectiveness of exorcisms, but there is evidence for the placebo effect. As such, exorcism offered help and relief to people who had no other options back in the day.

THE RENAISSANCE

After the long static period of the Middle Ages, interest in Greek and Roman ideas came back into fashion after they were deemed sufficiently retro-chic, and the Renaissance was thus born! And more than ever before, people of this era were like, "Let's crack open some skulls and dig into this thing, eh?"

A True Renaissance Man

Although he was better known for his artistic and technological genius, Leonardo da Vinci was obsessed with finding the seat of the human soul. Unfortunately for him, human dissections were illegal, so he paid a grave robber to pass him rotting bodies in the middle of the night. It turns out that Leo's artistic skills came in really handy, as he drew some fantastic images of the brain, nerves, and optic chiasm, where he thought the soul resided. Unfortunately, Da Vinci kept his drawings secret since they were illegal, and they weren't discovered until the 1800s. Random fun fact: Da Vinci did research on live frogs, but became disturbed by killing them, so he became a vegetarian. In fact, he loved animals so much that he would supposedly buy birds at the market and let them go.

$#@% You, Galen

If you remember one name from the Renaissance, it should be Andreas Vesalius. His dissections and discoveries took the world by storm. And he *hated* Galen, whose teachings were still held as gospel over a thousand years later. Vesalius knewbetter—because he actually used human bodies. The town he lived in had loose laws, allowing him to conduct public dissections and hire top-notch artists to draw for him. Thanks to his careful eye and detailed analyses, Vesalius identified multiple brain structures, including the corpus callosum.

Vesalius got some pretty harsh criticism for his work and for how he treated dear ol' Galen's

legacy. Vesalius didn't take that too well, so after he published his masterpiece on human anatomy, he burned the rest of his unfinished work, left medical school, and lived the rest of his life as a standard physician.

Whew! Our understanding of the brain went from being a useless thing we threw away to being the home of human consciousness! But no time to waste—let's skip ahead a few centuries.

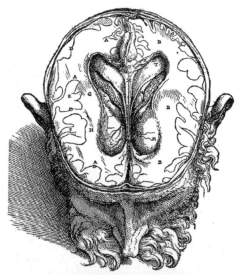

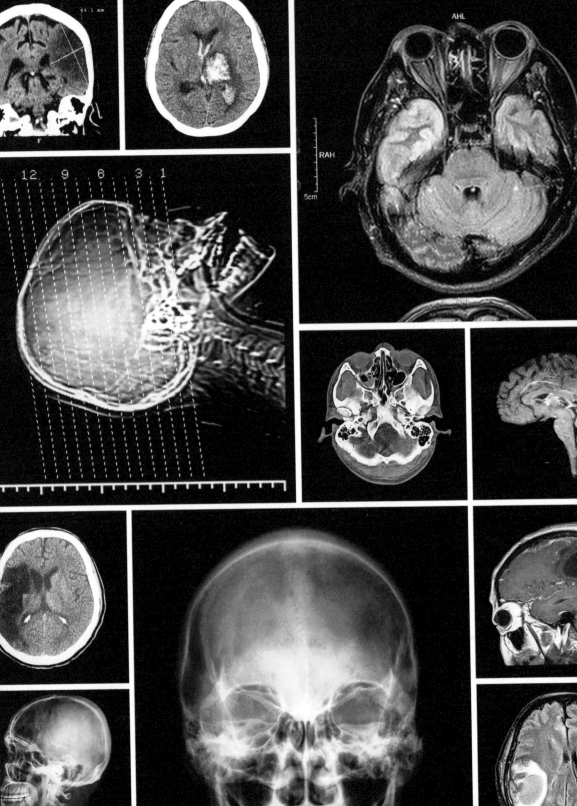

Chapter 3

LESSONS FROM SOME BANGED-UP BRAINS

Back in the day, studying the living brain was pretty difficult. After all, you couldn't just cut open someone's head and poke around in there! Or . . . you could, but usually it didn't end well.

As we've learned, early scientists weren't too sure what to make of the brain. Even if it wasn't made of sperm, it was a weird, gooey, mushy pile of . . . something. Even after technological advances made it easier for researchers to look at the individual cells in the brain, (more on that later) it still wasn't clear exactly how it all worked.

Early neuroscientists had to get pretty resourceful. And while it was generally frowned upon to just go around poking holes in peoples' heads, they could look at the holes that other people already had in their heads, as well as other (often less dramatic) brain injuries.

Long before modern brain science was even a thing, we already knew that head injuries could have seriously life-changing effects—losing the ability to make any new memories, developing a completely new personality, having visions, and more. The silver lining is that in the early days, when we were just starting to figure out how the brain worked, scientists could study people with brain injuries to better understand the importance of different parts of the brain. And some of the things they learned are still helping us understand memory, personality, and language today.

PHINEAS GAGE

So . . . how *do* you go about studying holes in the brain? And how do you know for sure that the person's brain injury and their new behaviors are connected? Well, one of the easiest ways is to look at how a person behaves while they're alive—and if there's something odd going on, open up their skulls (after they've died, of course) and take a look. And what might tip you off that something odd is going on in the brain? Some cases are more subtle than others. And then there's the case of Phineas Gage—the guy who somehow survived having an iron pole shot through his head, and taught us a valuable lesson about the importance of the prefrontal cortex in the process.

He'd Been Working on the Railroad

Our story begins in the 1840s in rural America. Polk was president, the Liberty Bell had just cracked, and the Washington Monument was under construction. And young Phineas Gage was helping to build a railroad in Vermont. Phineas was a very responsible and hard-working man, well-liked by his co-workers and described as "the most efficient and capable" employee by his boss.

Phineas's job involved setting blasting powder into holes drilled in the rock and packing in sand on top of the explosive powder with his trusty tamping rod to help direct the explosion. This mixture, like most things with "blasting" in their name, could be pretty touchy so it was veeery important to pack all of those things into the ground in the correct order. Maybe you can guess where this is going.

Brains and Blasts Don't Mix

One afternoon, while packing down a charge with the tamping rod and probably thinking about his lunch break, Phineas set off an explosion that sent the massive iron rod—almost 2 inches in diameter—shooting through his left cheek and out the top of his head before landing 80 feet away, covered in blood and brains. Miraculously, Phineas did not die. Within minutes, he was sitting up and talking and, after being taken back to his hotel, he was able to walk up a flight of stairs to his room.

He appeared so normal that when the doctor, one Edward Williams, arrived on the scene, he didn't believe Phineas' account of what had happened. But then Phineas "got up and vomited a large quantity of blood; the effort of vomiting pressed out about half a teacupful of the brain, which fell upon the floor." The doctor then proceeded to stick his fingers into the holes in Phineas' skull and cheek (I guess to confirm that the hole went all the way through?) before finally getting him cleaned up and wrapping up the wound.

Within a month, Phineas was able to walk, talk, and live independently, seemingly as he had done before the accident. But as he recovered, it became apparent that he had a kind of Dr. Jekyll/Mr. Hyde thing going on. After the accident, his personality completely changed. Suddenly, he was "capricious" and "irreverent," swore like a sailor, and couldn't be trusted to follow through on pretty much anything.

He soon lost his job and took off for a life on the road, tamping iron in hand, making a living as a medical miracle on display at the circus. Eventually, he wound up sick in San Francisco and living with his family, where he died twelve years after the accident.

A Hole in the Head

Unfortunately, by the time of his death Phineas was relatively obscure and no autopsy was conducted, so we're not totally sure what the damage to his brain actually looked like. A medical doctor named Harlow eventually recovered Phineas's skull several years after his death and inferred that he had probably suffered catastrophic damage to his left frontal lobe. The fact that "Gage was no longer Gage" after his injury led Harlow to speculate that the frontal lobe might be critical for decision-making, rational behavior, and personality.

Other researchers at the time thought this hypothesis was ridiculous, but almost two hundred years and many more brain injuries later, it's pretty clear that Harlow was onto something. We now know that the prefrontal cortex is vital for most of our higher-order cognitive functions, including social and moral reasoning, self-awareness and—you've guessed it—personality.

So it seems pretty clear that Phineas's dramatic personality change was a direct result of the damage to his frontal lobe. Or maybe his personality change was less about the brain injury itself and more about having a meter-long spike sent through his head. That would mess us up, too.

BRAIN INJURIES CAUSE *WHAT*?

Our brains are so complex (and vital to, well, everything) that it's not surprising how many different ways a brain injury may manifest, from the horrific to the . . . unusual.

Becoming a Genius
When people think "savant," they usually think of *Rain Man*, but he wasn't the only genius in the business. We're not sure exactly what causes it, but about half the time savantism is acquired due to injury or disease. Professor Allan Snyder believes that savantism is caused by our higher cognitive processes blocking our ability to access most of the information we have stored, and has demonstrated that blocking brain activity in the left anterior temporal lobe can artificially induce some similar phenomena.

Acquiring an Accent
Damage to certain areas of the brain associated with speech processing, including the cerebellum, can cause Foreign Accent Syndrome, where a person suddenly seems to have a "new accent" after a stroke or head injury. Unfortunately, despite miraculous claims of, say, a "Croatian [who] suddenly speaks fluent German," the condition isn't linked to any actual changes in linguistic ability. It's actually just a result of a person having a hard time pronouncing certain word sounds correctly.

Getting a Bit Mixed Up
Neurologist Oliver Sacks famously described the case of "the man who mistook his wife for a hat"—due to a condition called visual agnosia, the man couldn't recognize the faces of other people. This condition can be caused by damage to the pathway between the occipital lobe, where visual information is processed, and the parietal and temporal lobes, where that information is connected to memory and meaning.

PATIENT TAN

Phineas Gage's brain injury was . . . unusually dramatic. Not so dramatic are a whole class of serious, even lethal brain injuries that can occur almost stealthily. Some are caused by a person's own neurological condition, such as epilepsy, or after a stroke or concussion, resulting in areas of permanent brain damage, called lesions. The effects may be more subtle, or increase over time, but they can still cause all kinds of problems—like completely erasing your ability to say more than one word.

Call Me Tan. Tan Tan.

People get nicknames a lot of ways—a shortened version of their given name or an embarrassing joke from childhood. For French epilepsy patient Louis Victor Leborgne, the source of his nickname was simple: he was called "Tan" because "Tan" was the only word he was still able to say. At about the age of 30, he entered the Bicêtre hospital in Paris seeking treatment for what he and his family hoped was a temporary condition. After all, Leborgne had been successfully coping with epilepsy for years, so they weren't too worried when he lost all ability to speak . . . except for the word "tan."

The odd thing about Leborgne was that aside from not being able to say much, he seemed pretty normal. He appeared to understand everything that was said to him perfectly, and was able to accurately follow instructions. He would earnestly engage in conversation, albeit in a rather limited way on his end, and worked hard to communicate with those around him. Unfortunately, Leborgne's loss of words was only the beginning; over time, he began to experience paralysis and loss of vision, followed eventually by cognitive difficulties.

Bumps on the Brain

Around that time, Dr. Paul Broca was chilling in Paris with his psychology buds, arguing about whether or not different traits were localized to specific regions of the brain. Some of this group bought into the idea of phrenology, the theory that not only do functions like language and memory have specific locations in the brain, so too do things like intelligence and spirituality. Broca and his colleagues got so caught up in these arguments that they started looking for patients whose conditions might help settle the debate.

In Leborgne, Broca found such a patient. Broca immediately realized how important this case might be, though Leborgne was near death from gangrene. After the man's death, Broca performed an autopsy and found a lesion on the frontal lobe of his left cerebral hemisphere, just above his left eye. Broca called Leborgne's loss of language an "aphasia," from the Greek for "unable to speak." Today, this area of the brain is known as "Broca's area." Damage to this particular region means that a patient can still comprehend and respond to speech normally, with no other cognitive deficits, but is unable to produce words or sentences in a meaningful way.

Not Much of a Talker

Patient Tan, and other lesion patients who also experienced language deficits, essentially revealed to scientists the phenomenon of speech localization. This was some of the earliest evidence for cerebral localization, the once-controversial notion that different parts of the brain have specific functions. Broca's area actually has something of an equal and opposite number: when a patient has damage to the nearby region known as Wernicke's area, they can talk a lot, but usually only speak in gibberish, and don't seem to be able to fully understand what others are saying to them.

He might not have said much, but Leborgne's brain lesion and the subsequent research by Broca and others revolutionized our understanding of human language, and we're still making new discoveries today, over 150 years later.

What Broca Got Right

It is absolutely the case that damage to certain specific brain regions can have dramatic effects on specific traits and behaviors, because some things are localized to specific regions. As Broca noted, damage to the left temporal lobe can, in most cases, lead to difficulties with forming language ("most cases" because in fact not everyone has language localized to the left side; some people have it on the right, and it's more common to have it localized to the right side if you're left handed! And now, back to our regularly scheduled sidebar). In fact, the whole temporal lobe (on both sides) is critical for processing the sounds we hear. Ironically, the occipital lobe at the very back of the brain is responsible for processing all the visual information that's relayed from your eyes on the front of your head. As we learned from Phineas Gage, the frontal lobe is important for decision making, emotional regulation, and personality. Other brain regions are specialized to process complex movements, somatosensory (the sense of touch and balance) information, and even the formation and storage of new memories, as we'll learn on the next page.

What Broca Got Wrong

Broca's discovery of the area that now bears his name was born out of some pretty racist ideas that were common in the 1800's. He believed that people of different races were actually evolved from different species, and bought into the idea that the size and shape of certain physical traits, like the width of your jaw or bumps on your skull, were predictors of intelligence. It's probably no surprise that the traits traditionally associated with white Europeans were the ones that he claimed were linked to intelligence. This kind of unscientific logic got him interested in looking at brain size and ultimately whether or not language was linked to a particular brain region, which eventually led him to Leborgne. So even though he was right that certain brain regions can be important for specific behaviors, he was definitely wrong in thinking that you could tell how smart someone was by the width of their chin—or the size of their prefrontal cortex. It's not the size of the bump, it's the firing of the wiring!

HENRY MOLAISON (HM)

Sometimes brain lesions aren't caused by a direct injury. In fact, before we fully understood the roles of certain brain regions, doctors sometimes deliberately cut them out just for funsies (see: lobotomies). And sometimes patients seeking treatment for other conditions just got unlucky, like the unfortunate Henry Molaison, or H.M. as he was known to the medical establishment.

be to remove the affected tissue. Unfortunately, it was the 1950s, and we didn't fully understand exactly what those parts of the brain do. Still, the surgery went smoothly and without any surprises—until Henry woke up.

You ever see *Finding Nemo*? Henry was basically like Dory. After the surgery, Henry could still remember his name and his childhood, but he had severe anterograde amnesia— he couldn't form any new memories.

Finding Henry

Molaison was only a little boy when he had a bicycle accident that left him with severe epilepsy. Unfortunately for him, his seizures became worse as he grew older; by the time he was 26 years old they were so frequent and so intense that he was unable to do pretty much anything. And when seizures get that bad, one of the best treatments is to simply remove the parts of the brain that are causing them.

Luckily for Henry, his doctors were able to pinpoint the origin of his seizures to his left and right medial temporal lobes: the hippocampus, amygdala, and part of the entorhinal cortex. The doctors decided the best course of action would

Forever Young

This was really, really bad for Henry, but really, really good for the medical establishment. For the rest of his life, Henry was the subject of intense psychological and neurological study by dozens of doctors who sought to understand the long-term effects of his brain surgery.

By working with Henry, these scientists learned important things about the role of the hippocampus and the entorhinal cortex in memory formation. Henry only had a deficit in his long-term memory; he was still able to form habits, like finding his way to the hospital dining room, and perform working memory tasks, like repeating a string of numbers. This showed that while the hippocampus and nearby regions were clearly critical for forming new explicit memories—memories of facts and events—they didn't seem to be as important for forming implicit memories—the "unconscious" kind of memory that lets us do things like play an instrument or drive a car. And even though Henry experienced serious explicit memory deficits, his brain was sometimes able to find workarounds. For example, he could sometimes modify existing memories by "updating" memories he already had, and could draw a map of the house he lived in despite not being able to actually consciously remember its layout.

The study of Henry's learning abilities helped clarify the connections and differences between types of memory. It is a strange irony that, because of his brain lesion, he never had any idea how famous he was. He spent the rest of his life in a hospital, waking up each morning thinking he was still 26. How's that for forever young?

Life After Death

Henry died in 2008, at the age of 82. Even though doctors were able to image his brain using MRI technology while he was alive, scientists were grateful to have the opportunity to preserve his brain after his death. It was frozen and shipped to UC San Diego, where Alie completed her PhD, to be sectioned—cut into 2,403 *very* thin slices like deli meat, using a super sharp knife, in a process that was live-streamed for over 53 hours. His brain was then scanned and digitally reconstructed, so scientists around the globe could get a closer look at what, exactly, had happened to Henry's poor brain.

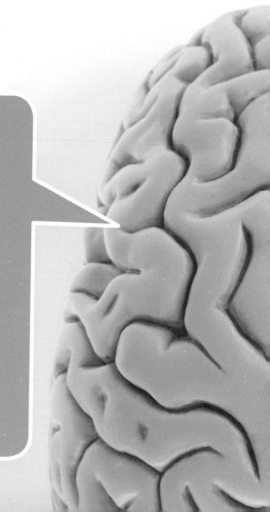

HEMISPHERECTOMIES: WHEN HALF A BRAIN IS BETTER THAN ONE

While Henry Molaison suffered severe long-term consequences after his surgery, not everyone who undergoes brain surgery to treat epilepsy ends up in such dire straits. In fact, in cases of severe epilepsy, surgical excision is still the most effective option for many people. In adults, these surgeries usually involve having the patient be awake and talking to doctors during the procedure, so the surgeons can keep track of how the patient is responding and make sure they're not removing any critically important brain tissue. But as for kids—well, kids' brains are still developing well into their teens, so there's a lot of space for the brain to adapt after an injury. So in very severe cases of epilepsy in very young children, doctors sometimes perform a dramatic surgery called a hemispherectomy, where they disconnect, disable, or remove an entire half of the brain. It seems like that should be pretty disabling, but in kids who are very young (under the age of 5), they actually end up pretty much normal. Or, you know. As normal as any of us ever are. The remaining half of their brain is able to compensate for the missing half by creating a lot of extra new connections to take care of all of that processing.

WHAT'S YOUR DAMAGE?

While the size of specific brain regions isn't associated with ability, there are still a lot of reasons why you don't want to go concussing yourself left and right. All these brain regions play pretty important roles in our lives, and damaging them can cause everything from a funny accent right up to . . . well, death. The diagram on this page tells you what certain brain regions control, and what happens if you damage them.

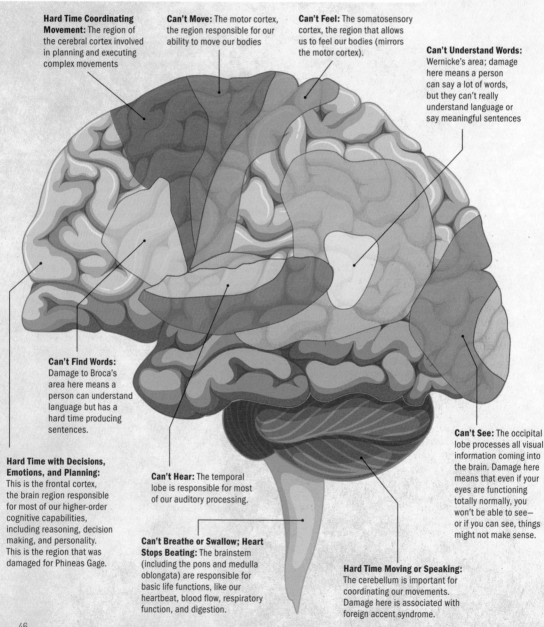

Hard Time Coordinating Movement: The region of the cerebral cortex involved in planning and executing complex movements

Can't Move: The motor cortex, the region responsible for our ability to move our bodies

Can't Feel: The somatosensory cortex, the region that allows us to feel our bodies (mirrors the motor cortex).

Can't Understand Words: Wernicke's area; damage here means a person can say a lot of words, but they can't really understand language or say meaningful sentences

Can't Find Words: Damage to Broca's area here means a person can understand language but has a hard time producing sentences.

Hard Time with Decisions, Emotions, and Planning: This is the frontal cortex, the brain region responsible for most of our higher-order cognitive capabilities, including reasoning, decision making, and personality. This is the region that was damaged for Phineas Gage.

Can't Hear: The temporal lobe is responsible for most of our auditory processing.

Can't Breathe or Swallow; Heart Stops Beating: The brainstem (including the pons and medulla oblongata) are responsible for basic life functions, like our heartbeat, blood flow, respiratory function, and digestion.

Hard Time Moving or Speaking: The cerebellum is important for coordinating our movements. Damage here is associated with foreign accent syndrome.

Can't See: The occipital lobe processes all visual information coming into the brain. Damage here means that even if your eyes are functioning totally normally, you won't be able to see— or if you can see, things might not make sense.

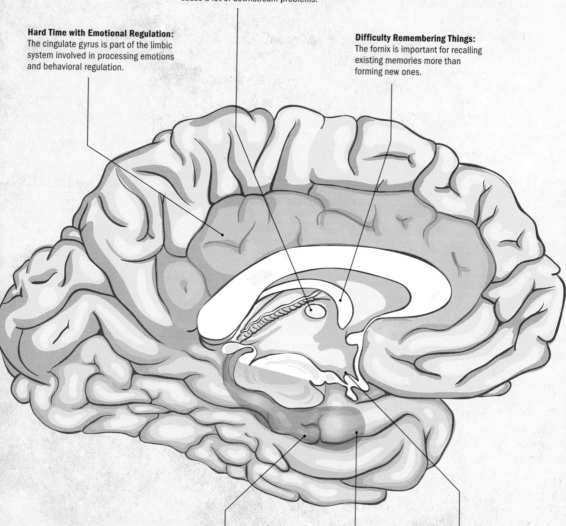

Sensory Changes, Movement Problems, Maybe Coma: The thalamus is a pretty critical relay center in the brain, so damage here can cause a lot of downstream problems.

Hard Time with Emotional Regulation: The cingulate gyrus is part of the limbic system involved in processing emotions and behavioral regulation.

Difficulty Remembering Things: The fornix is important for recalling existing memories more than forming new ones.

Can't Make Memories: that pesky hippocampus again. Without this seahorse-shaped structure, we can't form new memories.

No Fear: The amygdala is associated with risk taking, anxiety, and fear. Damage to the amygdala increases risky behavior and is associated with gambling.

Problems with Hormones: The hypothalamus is regulates our body's hormones, which affect things like mood, body temperature, growth, hydration, sleep cycles, and milk production.

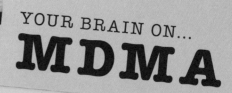

YOUR BRAIN ON...
MDMA

What It Is: 3,4-Methylene-dioxy-methamphetamine

What Type of Drug It Is: empathogen-entactogen and stimulant

What It Does: Let's start with what it doesn't do—turn your brain into Swiss cheese. I'd never heard of this myth (I guess I should have paid more attention in D.A.R.E.) but our editor told us about an anti-drug campaign that claimed MDMA causes "holes in your brain." Apparently this stems from the U.S. DEA's misrepresentation of a study in the 90's that showed that *after* taking MDMA, there are lower-than-normal levels of serotonin throughout the brain. Now, that's a real thing we already knew, and has nothing to do with holes in the brain. MDMA makes the user feel more physically stimulated and awake, while promoting sociability and enhancing physical sensations, especially touch.

How It Does This: MDMA acts by forcing your neurons to release as much of the neurotransmitter serotonin as possible, as well as oxytocin, norepinephrine, and dopamine, flooding the brain with these feel-good chemicals. MDMA also prevents the brain from cleaning serotonin out of the synapses so it can keep stimulating neurons, which keeps the high going and makes you feel euphoric and energetic.

What the Risks Are: All that excessive serotonin floating around can have negative effects on its own—too much serotonin in the short term can cause serotonin syndrome, which can lead to seizures and other neurological problems. While dumping out all that serotonin can leave you feeling kind of blue for a few days, over time, your body produces more serotonin and things mostly go back to normal. Frequent, heavy MDMA usage is associated with other problems, like shrinkage of some brain regions and possible memory impairments. But MDMA is also considered by some to be a miracle for its ability to encourage empathy and trust, and it's currently being explored for its potential to treat psychological conditions like PTSD—more on that later.

THAT'S GONNA LEAVE A MARK

There are all kinds of other things that can leave marks on your brain (literally. Like, not just in your memory, I mean). For example, some folks are born with "birthmarks" on their brain, thanks to a condition called Sturges-Weber Syndrome. It's visible at birth in the form of a port wine birthmark on the face, a sign of the abnormal growth of blood vessels in that area. Inside the brain itself, this can lead to problems with blood flow, which can in turn trigger everything from seizures and migraines to intellectual disability, and can sometimes lead to blood clots and stroke.

Even without any birthmarks, just the act of being born comes with a risk of brain injury. Birth-associated brain injuries include everything from brain cell death due to oxygen deprivation to nerve damage during a challenging delivery to

brain bleeds, skull fractures, and brain damage after the use of forceps to aid in delivery. Most of these problems stem from the fact that our big ol' brains usually get forced through a little ol' vagina on our way into the world. Thanks a lot, evolution!

Not everyone who's missing part of their brain sees significant changes in behavior or function. In fact, some people missing parts of their brain don't even know it. There have been some really incredible case studies on this, like the woman born without a cerebellum who, aside from having some balance issues, was pretty normal—and another where a man missing nearly 90 percent of his brain due to fluid buildup only experienced "mild leg weakness," but was otherwise a normal dude with a normal family and a normal job.

48

A WHOLE LOTTA NOPE:

LOBOTOMIES

Lobotomies were all the rage in the 1940's, hailed as a "miracle cure" for all kinds of mental illnesses and psychiatric breakdowns. But since performing a lobotomy essentially entailed jamming an ice pick into someone's skull and swiping it around to destroy part of the prefrontal cortex, the fad passed pretty quickly.

This procedure, initially known as a leucotomy, was developed by Portuguese scientist Antonio Egaz Moniz, based on research showing that destroying a chimpanzee's frontal lobes could subdue its aggressive behavior.

A Quick Stabbing Pain

In the early 1940's, American doctors James Watts and Walter Freeman developed a similar surgery, first using alcohol injections to directly kill brain tissue, and later using a modified ice pick to go in through the eye socket and cut up the brain. Watts and Freeman reported that over 60 percent of their patients were "improved" after surgery, though they did note (as did Moniz, funnily enough) that after surgery the patient's behavior frequently became more blunt and disinhibited. Oh, and . . . one in seven patients just straight up died. Which is a pretty serious side effect.

A surgery like the lobotomy was super appealing for many, mostly since it was really difficult for most people—especially if they weren't white or wealthy—to get any kind of mental-healthcare. The only option for many was institutionalization, but it wasn't great. Hospitals were overcrowded and patients were often kept isolated and physically restrained. In the face of such conditions, it's not hard to see why turning a difficult or dangerous patient into a more docile one would be so appealing, even if it came with the risk of turning them into a vegetable.

Women and Children First

If you're on the fence about this whole lobotomy thing, one major red flag is how frequently they were performed on people already pathologized by society as "different," meaning they had behaviors or beliefs that didn't adhere to the era's strict social norm. In fact, a lot of lobotomies were performed on women, who were expected to recover relatively easily and go back to their household duties.

If you think it can't get worse than this, it does. Lobotomies were often done to children as well, to make them "more docile." In fact, Freeman himself thought that African Americans, and particularly African American women, were the best patients for lobotomy because they were the "most likely to have a supportive family at home to provide them with lifelong care."

This makes it all the more upsetting that the procedure was so popular that Antonio Egaz Moniz actually won the Nobel Prize in Physiology or Medicine in 1949 because of his "discovery of the therapeutic value of leucotomy in certain psychoses." Yeah. Some pretty big mistakes were made.

Drugs Save the Day

Eventually, concerns about the method of performing a lobotomy as well as growing discomfort with the negative outcomes of the procedure made many doctors uneasy. Then, in 1950, a new drug called chlorpromazine was invented—an antipsychotic drug that was hailed as a miracle for treating schizophrenia. More and more doctors began to point out that the lobotomy had essentially no scientific basis. By the 1970s it was outlawed in many countries and U.S. states, and has since been determined to be appropriate only in extremely rare cases.

AMNESIA

How well does Hollywood do with this complicated neurological condition? Here's a brief, yet totally scientific, survey.

The Majestic (2001)

Peter Appleton (played by Jim Carey) is a 1950s Hollywood screenwriter who's accused of being a Communist. On a bender, he drives his car off a bridge and washes up on a beach, with no memory of who he was. Residents of a nearby town think he's a local veteran returned from the war, but Appleton realizes the truth when he sees his name on the big screen and ends up fighting against the accusations of communism, feeling inspired by the actual war hero he had been mistaken for.

Depiction of Amnesia: ★★☆☆☆
While retrograde amnesia can and does happen, it's pretty unlikely that you'd suddenly "snap back" to yourself after one sight of your name.

Plot: ★★★☆☆ If you want a sentimental film with Jim Carey playing a serious role, *The Majestic* is for you. It's not "great cinema," but it's sweet and simple.

50 First Dates (2004)

Henry (Adam Sandler) creepily falls in love with Lucy (Drew Barrymore) and woos her day after day, even though she forgets him every night.

Depiction of Amnesia: ★★★☆☆
The depiction of anterograde amnesia is okay, but Lucy probably wouldn't remember things for very long at all—making the romantic scenes preeeeetty uncomfortable.

Plot: ★★★☆☆ I get that she seems great but it's still weird to fall in love with someone who can't ever remember you!

Finding Nemo (2003)

Nemo (Alexander Gould, as a fish) runs away from home because of his overprotective dad Marlin (Albert Brooks, also a fish). In the search for his son, Marlin meets a memory-challenged Dory (Ellen Degeneres—also a fish, but blue this time) who manages to remember things when it counts the most.

Depiction of Amnesia: ★★★☆☆
Dory's anterograde amnesia is pretty accurately portrayed, but it's sort of deus ex machina to have her suddenly be able to remember "P. Sherman, 42 Wallaby Way, Sydney" when it's such a crucial plot point.

Plot: ★★★★☆ Pixar can do no wrong, in my opinion, but jeez, Albert, don't be so overbearing!

The Bourne Identity (2002)

Matt Damon is very ripped and good at killing people but can't remember why. Turns out he was a government-sponsored murder machine, and now they're trying to murder him.

Depiction of Amnesia: ★★★★☆ Pretty good depiction of retrograde amnesia (inability to remember the past) but it's hard to tell what triggered the amnesia in the first place, so hard to know how accurate it was.

Plot: ★★★★★ Very gritty; peak hot Matt Damon.

Memento (2000)

Guy Pearce has a lot of tattoos for some reason, and is trying to find the guy who killed his wife. Lots of people take advantage of his amnesia in the process.

Depiction of Amnesia: ★★★★★ Guy can't remember anything for more than a few minutes, hence the tattoos and frequent polaroid photos. Actually one of the most accurate depictions of anterograde amnesia on the screen.

Plot: ★★★☆☆ I know lots of people really liked this movie, but I thought it felt like Nolan was trying too hard.

SORRY TO BREAK IT TO YOU, BUT . . .

Losing consciousness for more than a few seconds is usually REALLY bad. Like frequently fatal, need-to-go-to-a-hospital-right-now bad . . . not "shake it off and keep the plot going" bad.

Getting shot in the head doesn't always kill you immediately. But it's definitely going to leave your body twitching a lot more than the movies depict regardless.

Getting slammed into the ground, a wall, or a car's windshield is probably going to just like . . . destroy your brain. For real.

There's no such thing as "unlocking your brain with drugs." Your whole brain is working all the time and drugs definitely won't let you bend reality. Sorry, *Limitless* and *Lucy* fans.

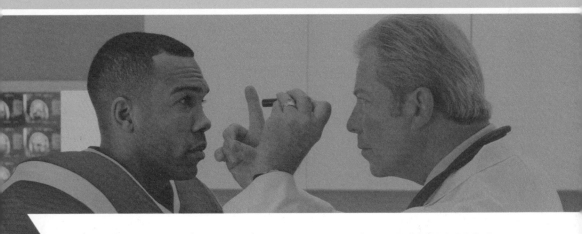

CAN A CONCUSSION GIVE YOU AMNESIA?

Yes. But a second bonk on the head probably won't fix it (despite what movie tropes would have us believe). Amnesia can be caused by anything that damages the brain, and that definitely includes concussions. It's not uncommon to have some memory troubles immediately following a concussion, thanks to a phenomenon known as diaschisis, where your fragile post-injury neurons have a hard time talking to one another properly. When the injury isn't too severe, often the amnesia clears up over time—but hitting your head again definitely won't help, since all you'd be doing is injuring even more of your brain.

RIGHT BRAIN VS LEFT BRAIN

Are you left brained or right brained? There are quizzes abound to tell you which hemisphere "dominates" your personality, as well as tailored life advice, career coaching, and study tips designed to help you succeed no matter which brain you're using. Such a shame it doesn't actually work like that.

The Misconception
It makes sense that people started to think that one side of your brain could dominate your behavior and personality, because some traits are associated with one side of the brain or the other. It's not too wild to think that the left side of your brain might be more logical and analytical, while the right brain might be more creative and emotional. All of this stems from the misconception that one half of your brain is more in charge than the other, which isn't really true.

Why Did We Think That?
The reality is that, each half of your brain controls the other half of your body—so the right side of your brain controls the left side, and vice versa. In a healthy, ordinary brain, the two sides of the brain are connected via an information superhighway known as the corpus callosum, passing information back and forth to keep your whole body coordinated.

There are some rare surgeries to treat conditions like epilepsy which require doctors to cut through the corpus callosum. This has a whole lot of weird effects, and because some behaviors are based in one hemisphere or the other, some of those effects can make it seem like certain personality traits are localized to one side of the brain.

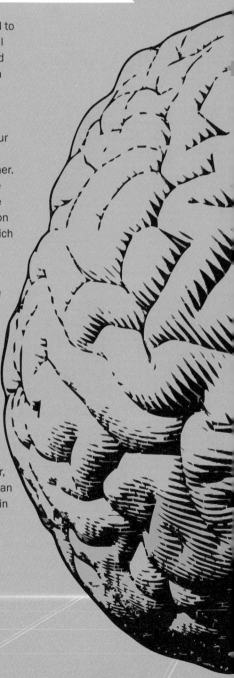

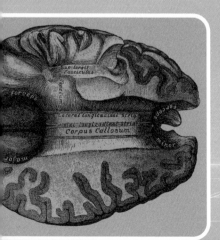

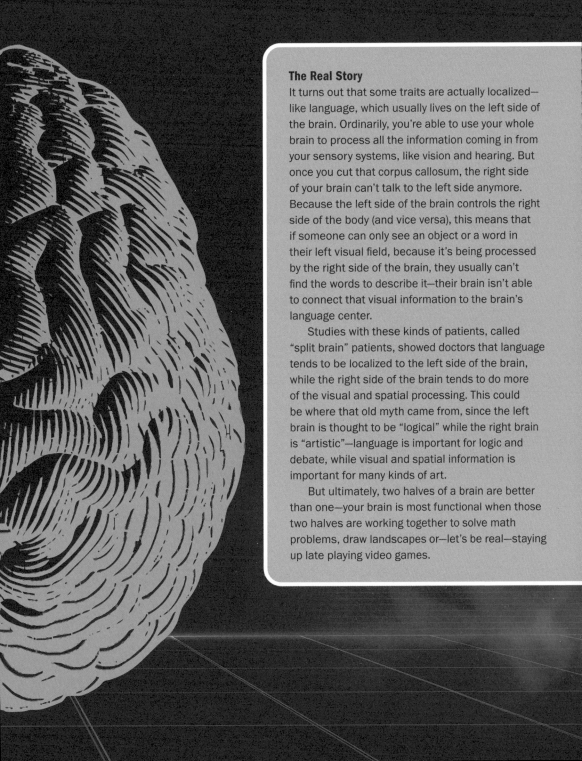

The Real Story

It turns out that some traits are actually localized—like language, which usually lives on the left side of the brain. Ordinarily, you're able to use your whole brain to process all the information coming in from your sensory systems, like vision and hearing. But once you cut that corpus callosum, the right side of your brain can't talk to the left side anymore. Because the left side of the brain controls the right side of the body (and vice versa), this means that if someone can only see an object or a word in their left visual field, because it's being processed by the right side of the brain, they usually can't find the words to describe it—their brain isn't able to connect that visual information to the brain's language center.

Studies with these kinds of patients, called "split brain" patients, showed doctors that language tends to be localized to the left side of the brain, while the right side of the brain tends to do more of the visual and spatial processing. This could be where that old myth came from, since the left brain is thought to be "logical" while the right brain is "artistic"—language is important for logic and debate, while visual and spatial information is important for many kinds of art.

But ultimately, two halves of a brain are better than one—your brain is most functional when those two halves are working together to solve math problems, draw landscapes or—let's be real—staying up late playing video games.

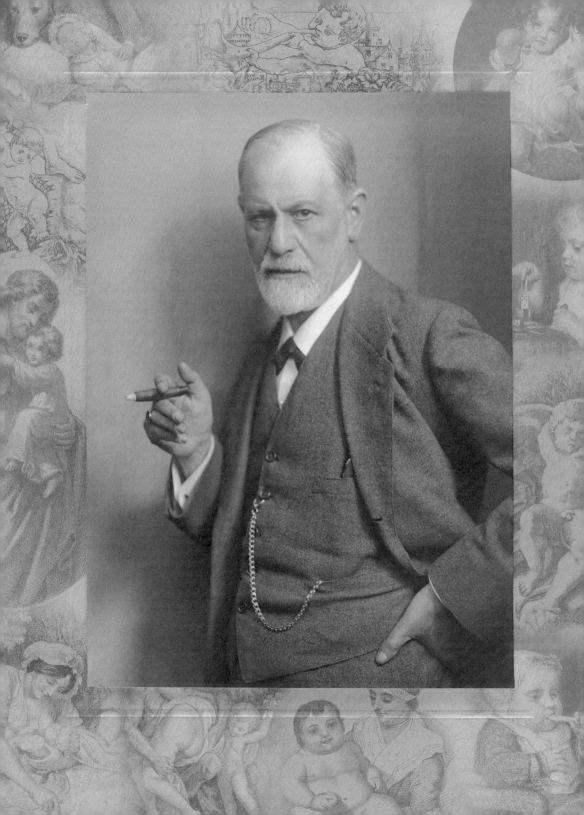

Chapter 4

WAIT . . . SO, KIDS AREN'T JUST TINY ADULTS?

Sigmund Freud had some weird thoughts about babies. Like, super weird. Let's just say that our understanding of child development was in its infancy at the time.

Growing up is probably the weirdest, most embarrassing, most magnificent thing we go through in life. Think about it: We morph from helpless infants with giant heads into energetic children with boundless imaginations, then into gangly teens with identity issues, and finally into these eccentric adults full of unique hang-ups we just can't seem to get over. And those early years appear to have the largest impact on the rest of your existence. There's no escaping it— you are who you are today, in great part, thanks to your upbringing. So logically, if you want to

understand yourself now, then you need to look backwards, right?

. . . Okay, I can't actually hear you because this is a book, but if you answered "yes," then you're in good company. For the last century, psychologists have tried to explain how children develop from, say, an expensive crying machine into a full-fledged grown-up who enjoys reading a book about their brain. So let's explore a few of these theories of development, starting off with the man, the myth, the legend, the stone-cold weirdo: Sigmund Freud.

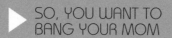

Sigmund Freud is one of the most controversial figures in psychology. On one hand, he's a genius. Many consider him the father of modern psychotherapy, he founded psychoanalysis, and he expanded psychology's reputation among academics and the public alike. And even today, his influence lives on through psychological terms he coined that we still use in everyday language.

On the other hand, from our modern point of view, Freud was kind of a quack. Many of his theories have been debunked because they had no supporting evidence. Also, he appears to have been a terrible chauvinist who, oh yeah, thought our sex organs connected directly to our nose. And he really loved cocaine . . . a lot.

Prolific in Every Way

Regardless of whether you love or hate him, Freud was a very interesting guy. He learned eight different languages, escaped Nazi capture and certain death, was nominated for the Nobel Prize thirteen times, became widely known for his fascination with sex, and (perhaps unsurprisingly) had six children with his wife, Martha Bernays. It was during his time as a father that Freud noticed that, at particular ages, his children became fixated on certain parts of their bodies. Freud being Freud, he figured that this must be for some sexual reason that might affect you later in life. Now, you might be saying to yourself, "It's kind of weird to sexualize infants." And you know what? You're right! But that didn't stop Freud, and he eventually developed his thoughts into his theory of Psychosexual Development.

Stage Fright

Freud divided up child development into five distinct stages: Oral, Anal, Phallic, Latency, and Genital. Each of the first stages is associated with an "erogenous zone" like the mouth, bowels, or genitalia . . . although it sort of seems like Freud got lazy after the third stage because he doesn't have any new body parts for the Latency or Genital stages. I'm not sure what the natural progression would be. What do you explore next? Feet, maybe? Clavicles? Anyway, Freud theorized that getting somehow mentally stuck at any of these stages would result in issues that would persist into adulthood.

TERMS FREUD INVENTED

Even if you know next to nothing about Freud, you still probably use all kinds of words and concepts he made up or popularized. Here's just a small sample.

Ego: The mediator in the mind between your instinctual desires and internalized morality. Of course, nowadays having a "big ego" means you're just a dick.

Death Wish: An unconscious desire for you or another person to die, often resulting in guilt and self-punishment. I dunno how much guilt BASE jumpers feel, though.

Defense Mechanisms: When your grandma brings up how you never call, this is how your mind protects you from feeling anxious or guilty. Geez, I'm just really busy . . .

Libido: The sexual energy that (according to Freud, anyway) drives all behavior . . . like driving you to a booty call.

Projection: Denying feelings about yourself, and dumping them on someone else. Pot, meet kettle.

Unconscious: The part of your mind that you're unaware of, but still influences actions and feelings. This is where Freudian Slips are born!

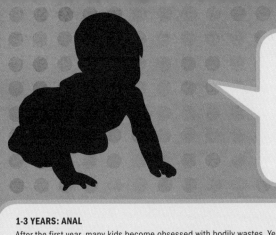

BIRTH-1 YEAR: ORAL

Freud believed that, in the oral stage, babies focus on the mouth as the source of pleasure. Infants receive gratification from breastfeeding and are only interested in their immediate needs. Freud thought that, if a baby struggles with weaning, they develop problems with trust and independence, becoming overly passive, immature, or optimistic. This leads to an "oral fixation" as adults; they may seek out activites that stimulate the mouth: smoking, gum-chewing, nail biting . . . and oral sex, of course.

1-3 YEARS: ANAL

After the first year, many kids become obsessed with bodily wastes. Yes, we've entered the anal stage, where children start to understand their parents' expectations and try to balance them with their own desire to poop wherever. With overly strict parents, Freud thought a toddler may become "anal retentive" and, as an adult, obsessed with neatness and order. If the parents are too lax, the toddler may become "anal expulsive" and be self-indulgent, reckless, defiant . . . and may be "into" feces. (You know what we mean. Sorry.)

3-6 YEARS: PHALLIC

Here, the child enters the phallic stage, perhaps the most bonkers of Freud's stages. Around this time, kids become aware of their bodies and gender differences—specifically, their genitals. Freud believed that this inherently changes the child's interactions with their parents, leading to the controversial Oedipus and Electra complexes. The Oedipus complex theorizes that, well . . . that every young boy wants to have sex with his mom and will actively compete with his dad for her attention. The Electra complex, meanwhile, also posits that girls want to have sex with their moms but, lacking the equipment, develop "penis envy" which somehow turns into wanting to have sex with their fathers. We didn't make this up! Freud thought a fixation at this stage would make girls grow into (sexually) dominant women and boys into (sigh, sexually, of course) aggressive men.

6-PUBERTY: LATENCY

For such a sex-obssessed dude, Freud actually laid off on the pre-puberty kids. Which is . . . something? After all that whirlwind of envy and gender drama, he figured nature would give us some time off until puberty. If only, right?

PUBERTY+: GENITAL

Yay, you're all grown up and nothing can ever mess with your head ever again. Okay, that's totally not true, but at least once you hit puberty Herr Doktor Freud stops trying to figure out why you're into the weird stuff you like. So that's . . . something.

Freud's psychosexual theory remained the predominant understanding of child development for over fifty years, until his daughter's pal Erik Erikson was like, "Whoa, bro, let's take the sex out."

Inauspicious Beginnings

Erik's mother came from a prominent Jewish family in Copenhagen; she fled to Germany after getting pregnant (it's not certain who the father was). There, she married and gave Erik his step-father's last name: Homburger. Erik learned all of this as a teenager, and didn't take it so well. Who was he, really? Feeling deceived and confused, he dropped out of school and roamed around Europe as a wandering artist. Eventually, he became an art tutor for a few wealthy families in Vienna . . . all of whom also just happened to receive psychoanalysis from Anna Freud, Sigmund's daughter.

One Little Suggestion

Erik went on to study psychoanalysis and specialized in working with children, which was a good fit—people always commented on how good he was with kids. But as Erik learned about Freud's theory of psychosexual development, he was . . . skeptical. To be clear, he was super into Freud's idea that everyone progresses through distinct stages and that getting stuck at any stage would mess you up for life. He just wasn't all about the wanting-to-bang-your-mom thing. (Which, yay for him.)

We Are Social Beings

Erik realized that there's more to growing up than just your relationship with your parents. We live in a society that shapes us all. From this realization, he developed a psychosocial theory of development that focuses on each individual's personal journey through their environment and the formation of their identity. It should come as no surprise that "identity" was a big focus, given Erik's history of confused feelings. After he moved to the United States, Erik changed his last name to Erikson to resolve his own "identity crisis," which is a term that he coined!

Five Stages? How About Eight!

Inspired by Freud's format, Erikson came up with eight developmental stages, extending beyond childhood, through adulthood, ending at your inevitable death. Each stage has a "task" that you either master or fail. Mastering these tasks is how someone becomes a successful, productive member of society. Failure at any task leads to feelings of inadequacy and, he postulated, some pretty gnarly consequences. Surprisingly, this theory still holds up, as it's one of the few that focuses on development throughout the entire lifespan.

Here's a nifty little reference chart for you. Note the explicit lack of sex in this one. Also, bonus, no poop!

Task	Age (years)	Key Question This Task Answers	Outcome of Mastery	Outcome of Failure
Trust vs. Mistrust	Birth - 18 months	Is my world safe?	Aww! You trust people. It must be thanks to your parents who gave you enough attention and love.	Booo! You mistrust people because the world is unpredictable and cruel. Hope you don't become an anxious wreck.
Autonomy vs Shame/Doubt	2 - 3	Can I do things by myself or do I always need to rely on others?	Wow! Look at how independent you are. You feel sure enough to do stuff like choose your own clothes.	Really? You clearly can't do this on your own. You should be ashamed. Have fun with your low self-esteem.
Initiative vs Guilt	3 - 5	Am I good or bad?	Holy smokes! You're so driven and confident. Look at how high up you are on that tree!	Hey! Get down from there! What makes you think you can do things without asking first. You need some serious inhibitions.
Industry vs Inferiority	6 - 11	How can I be good?	Incredible! You have such great friends and are doing well in school and soccer. You should be proud of yourself.	No! What is your problem? You have issues with other kids at school, you act out at home . . . Don't tell me this is an "inferiority complex."
Identity vs Role Confusion	12 - 18	Who am I and where am I going?	Look at you! You really seem to know yourself. You have such strong beliefs and values.	Ugh! You're always so apathetic. Stop doing what your parents want and figure out your future.
Intimacy vs Isolation	19 - 40	Am I loved and wanted?	Yay! I'm so happy you've found intimate, strong, loving relationships in your life.	Huh. So you're still single and miserable, eh? I bet you feel pretty isolated and lonely.
Generativity vs Stagnation	40 - 65	Will I provide something of real value?	Dang! You've got a meaningful career and you mentor kids? You're leaving a mark on the world!	Man. You are so shallow and self-centered. Can't you spend your time doing something to improve yourself?
Integrity vs Despair	65 - Death	Have I lived a full life?	Whew! You must feel pretty satisfied with what you accomplished in life. You are so wise.	Unfortunate. You wasted your life. Now you're just a bitter, depressed old person filled with regrets.

PIAGET ► TURNS OUT KIDS AREN'T JUST DUMB ADULTS

Jean Piaget was a pretty unusual child. His boundless fascination and keen perception of the world around him made him something of a scientist from a young age. By the age of 15, Piaget had already published multiple papers on shellfish, and went on to earn his doctorate in natural history at 22. But in the midst of studying clams and snails, Piaget decided to spend a semester studying under Carl Jung. (Just to tie the strands together, Jung had studied under Freud . . . until their working relationship fell apart due to, among other things, Jung not being sure that *everything* had to be about sex. Sensing a pattern?) Little did he know, his life path was about to significantly change.

Curious . . . Very Curious

His study with Jung seduced young Piaget away from a promising career in mollusks, and he spent the year after he graduated at the Sorbonne studying abnormal psychology. In Paris, young Piaget worked in psychologist Alfred Binet's laboratory, helping to develop the Binet-Simon Scale, one of the first standardized intelligence tests. But as he administered the test to hundreds of French kids, Piaget noticed a strange pattern. All of the younger children consistently got the same questions wrong. What the heck was going on?

Well, No. But Yes!

Feeling that same rush of excitement he used to get from contemplating a perfectly intact oyster shell, Piaget asked these children to explain the logic behind their incorrect responses. Surprisingly, he found that they actually had totally rational, intuitive explanations. See, up to this point in history, it was generally assumed that kids were just less competent thinkers than adults and got questions wrong because they were guessing the answer—and badly at that. However, Piaget found that children at different ages seemed to have totally different ways of thinking—their perspective shifted from an egocentric view as a baby to a more "sociocentric" view as they grew up and interacted with their peers. Giddy as a French schoolboy, Piaget tested his hypothesis and came up with his now-famous theory of cognitive development, which identified four stages that every child passes through. There are a number of ways to illustrate these phases, but probably the most fun is by looking at how to trick kids at each stage. Let's learn how to mess with kids on the next page, shall we?

SENSORIMOTOR STAGE (BIRTH-2 YEARS)

Babies explore the world through their senses and movement, which is why they put everything in their mouths.

How to trick them: An "object permanence" test. Just put a toy under a blanket and they'll think it disappeared! Ceased to exist! Now you can see why peek-a-boo is hilarious.

PRE-OPERATIONAL STAGE (2-7 YEARS)

Kids start to engage in imaginative play and can use symbols or images to represent other objects. But they're terrible at logic.

How to trick them: A "conservation" test. Pour water from a wide cup into a skinny cup and they'll think the skinny cup has more water! Or if they want eight fish sticks but there are four, just cut them in half and they'll be happy!

CONCRETE OPERATIONAL STAGE (7-11 YEARS)

Now that these kids are a little older, they're a bit more logical, but still struggle with abstract or hypothetical thinking.

How to trick them: A "deductive reasoning" test! Tell the child, "If you hit a glass with a feather, the glass will break. Johnny hit a glass with a feather. What happened?" Despite the setup, they'll say nothing happens! After all, feathers can't *actually* break glass. Very concrete thinking indeed.

FORMAL OPERATIONAL STAGE (12+)

At this point, kids have fully developed their thinking. They're hypothetical-problem-solving, abstract-thinking, deductive-reasoning, metacognitive machines! Thanks to this stage, teenagers start to think more about morals, ethical, political, and social issues that require that abstract thinking. Some might even become *gasp* philosophers.

How to trick them: Get them to apply to grad school.

VYGOTSKY ▶ IN SOVIET RUSSIA, SOCIETY PART OF YOU!

What's more impressive: Being one of the greatest psychologists in history, or living with low self-esteem in Soviet Russia, dying from tuberculosis at age 37, and still being one of the greatest psychologists in history? Not much is known about Lev Vygotsky's private life, in part due to his untimely demise. But he uncovered two important factors that influence our development: society and culture.

Everything Matters

Believe it or not, most psychologists poo-pooed the significance of society and culture. Freud, Erikson, and Piaget all generally believed that, no matter where you were born, you'd develop more-or-less the same as everybody else. And that makes sense if you're only studying white American or European kids. But Vygotsky was

like, "Nuh-uh, my dudes, you're missing a big chunk here." Vygotsky recognized how deeply your community could influence you while growing up. And contrary to popular theory at the time, Vygotsky believed that learning through social interaction and play leads to development of the brain, rather than the other way around. The language you speak, the neighborhood you live in, the culture you identify with . . . everything makes a difference in how you grow up.

Can You Give Me a Hand?

This idea led to Lev's greatest concept—and also an awesome name for a prog-rock band: the Zone of Proximal Development. This term describes tasks that a child (or adult) can accomplish with help. It exists between what the child already knows how to do and what is not possible for them to do yet. Unlike previous theories that propose a sequential path to development, Vygotsky realized that younger children can learn a new skill that is typically associated with older children if they receive guidance from a more knowledgeable instructor. For example, learning how to swim or read or solve puzzles. So essentially, the Zone of Proximal Development shows why teachers are so important and why they need to tailor their lessons for each student! Lev Vygotsky died before he could complete much of his work, but his ideas have lived on in classrooms around the world.

DO WE STILL USE THESE THEORIES TODAY?

Many of these theories were developed in the early 1900s when psychology was still a new field. Yet there's still no consensus on one theory of child development to rule them all—no one theory to bind them. That doesn't mean these theories are totally unchallenged, though. Much of Freud's psychosexual theory has zero supporting evidence. Erikson's theory was Eurocentric and didn't address how culture influences a child's progression. And Piaget's theory, though taught as gospel in psychology classes, gives the false impression that development happens in a sequential, fluid manner. They're still used today, but they are flawed creations.

DOES MOZART MAKE BABIES SMARTER?

A study published in *Nature* in 1993 found that college students performed better on spatial tasks right after listening to just ten seconds of a Mozart sonata. So if Mozart make adult smart, it make baby smart too, no? Turns out this one is a big ol' myth.

Music is generally really good for your brain, but the results of this study got blown way out of proportion. Like, the-state-of-Georgia-proposed-a-bill-that-gave-every-newborn-a-Mozart-CD-for-free out of proportion. But this phenomenon can be explained by something pretty simple: short-term arousal. If your brain wakes up and starts paying attention, it increases your short-term performance on tasks like playing a video game or doing a jigsaw puzzle. But you can also get the same effect by spending a few seconds sipping your coffee or reading this book. Unfortunately, there's no long-term effects on intelligence. Sorry, baby!

BOWLBY ► HE PROTEC, BUT HE ALSO ATTACH

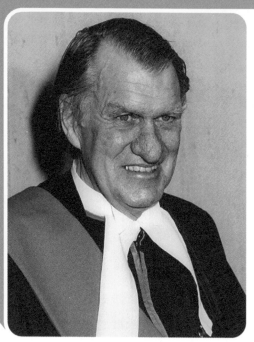

In the early 1900s, a well-to-do family in London had their fourth child, John. As was common at the time, all of the children were taken care of by a full-time nanny. Indeed, it was thought that having too much contact with their parents would lead children to become "spoiled." As such, John only saw his mother for an hour each day, after tea. How British. When John was four years old, his nanny left to find other work, which devastated poor little John. He grieved for her as though he had lost his own mother. And in a way, he had. John was sent away to boarding school a few years later and never regained that motherly love he needed. He felt . . . detached.

Please Don't Go

This young man, John Bowlby, was to become one of the world's most famous psychologists, and his boyhood experiences led him to develop his best-known idea: Attachment Theory. Bowlby was really interested in "bad" kids and how they became so delinquent. While working with children who steal, he realized that most of them had experienced some sort of extended separation from their parents before they turned five years old. Later, during World War II, he noticed how distressed children would become when they were separated from their mothers, even if there was someone else there to take care of them.

Bond . . . Brain Bond

Combining both of these observations, Bowlby concluded that everyone at birth starts to form a "lasting psychological connectedness" with the person who cares for them most. And that bond at an early age can seriously impact how you develop socially, emotionally, and cognitively. If that bond is healthy and strong, you will likely grow up to be a successful, well-adjusted adult. If it's weak or damaged, well then you're gonna have some issues, bud. Bowlby's theory of attachment opened up a whole new area of psychology and has changed how we approach parenting . . . like how you shouldn't only spend an hour with your kids after teatime.

DOES DAYCARE MESS UP ATTACHMENT?

It has become more and more common for parents of young children to be employed and working outside the home. As such, these kids spend time—sometimes a lot of time—in daycare facilities. Despite being separated from their parents, evidence shows that children placed in good daycares not only improve cognitively and socially, but also experience no negative impact to their healthy attachment. As long as they get warm, supportive, attentive care, it doesn't matter who they get it from. Just be careful! A low-quality daycare experience, like low-quality parenting, has the potential of messing up your child's attachment security.

AINSWORTH

Some folks believe that we develop attachments because it's an evolutionary advantage. In prehistoric times, our ancient ancestor parents were more likely to protect and take care of a helpless blob (aka baby) if that blob formed a strong bond with them. But of course, not all parents are good at parenting. In the 1970s, psychologist Mary Ainsworth realized that if the caregiver somehow damages trust in the parent-child relationship, the baby will learn maladaptive behaviors that help them get their needs met. Ainsworth noticed patterns in these behaviors and identified four distinct "attachment styles."

Secure Attachment

When caregivers are supportive and caring, the child expresses and receives love in an appropriate way: They develop healthy ways to cope with stress, and are more likely to grow up to be a confident, well-adjusted person . . . hopefully. About 70 percent of all children seem to have a secure attachment.

Anxious-Avoidant Attachment

These children will avoid or ignore their parents and seem emotionally distant. As an adult, this means relationships may not be that important to you. You might seek out a romantic relationship, but then lash out or end it if the other person gets too close.

Anxious-Ambivalent Attachment

Some children are often clingy and require constant reassurance. At the same time, they become resentful or shut down when the parent tries to comfort them. Adults with this attachment style may have low self-esteem and be overly dependent on their family or partner.

Disorganized Attachment

With this attachment style, a child is inconsistent with their needs and may be inconsolable at times because they do not know what they want. An disorganized attachment adult might have difficulty trusting others and may not even see themself as worthy of being close with others.

SECURE

ANXIOUS-AVOIDANT

ANXIOUS-AMBIVALENT

DISORGANIZED

MICAH'S CAT EXPERIMENT

Bill and Loki are both really good cats, but I often wonder whether we have a healthy bond. I'm an involved cat parent, so hopefully yes? I decided to use an experiment called the Secure Base Test to find out their attachment styles. You can try this at home with your pets! Here's how you do it:

Step 1: Take your pet into an unfamiliar room.

Step 2: Hang out with them for two minutes while they explore.

Step 3: Exit the room for two minutes, leaving your pet in the unfamiliar room. It is likely this will upset your pet.

Step 4: Re-enter the room and observe their behavior.

If your pet comes up to you for comfort and then continues exploring the room like Loki did, that means they're securely attached. If they stick to you like glue and keep needing comfort like Bill, that means they have an anxious-ambivalent attachment. Poor Bill! If your pet avoids you when you come back in the room, that means they have an anxious-avoidant attachment. And if they seem to have difficulty being comforted, like they come to you for help but then move away from you, they likely have disorganized attachment. Try it out! . . . If you can stand the fact that it might make your kitty cry.

A WHOLE LOTTA NOPE:
HARRY HARLOW'S MONKEYS

Some of what we know about childhood comes from some highly ethically unsound (read: really, really messed up) research that was conducted in our home state of Wisconsin by the infamous Harry Harlow. Harlow wanted to study what happens when a child is separated from their mother. Maybe he was deprived of hugs as a child; I don't know.

Poor Baby Monkeys
To test this, he separated young monkeys from their mothers after birth and placed them in a small cage that had two fake "mothers." One fake mom was made out of wire and equipped with a milk dispenser and the other mom was made out of soft cloth. To Harlow's surprise, baby monkeys spent most of their time clinging

to the cloth mom for comfort and would only go to the wire mom when they were hungry. This showed the importance of connection in relationships between a child and their parents. But Harlow was a sick monster, so he didn't stop at simple monkey neglect. He created other versions of the cloth mother that he called "evil mothers," each of which were equipped to torment the baby monkeys in various ways: The evil mothers would blast the babies with cold air, drench them with ice water, or literally stab them with retractable spikes. Harlow found that, despite being hurt by the cloth mom, the monkeys kept coming back for comfort—even until they were killed. Although this provided some insight into attachment to abusive parents, this experiment was extremely cruel and sadistic.

A Monster In Human Clothing

Harlow had a long history of horrific studies, including one about isolation in which he bound baby monkeys in a dark chamber for up to a year. He literally called it the "pit of despair." If they survived, these animals were psychologically disturbed and emotionally shattered. To what end? Harlow claimed it could provide a model for human depression, but when's the last time you were tied up in a dark room for a year? If it isn't clear yet, Harlow was a terrible person. Beyond his mistreatment of animals, he was also just . . . not very nice. He had ongoing extramarital affairs, treated his children coldly, and was a terrible alcoholic. (Our editor hopes that he's in Hell being tortured by monkeys for all of eternity. I wouldn't go that far, but I'm really glad we can't do these kinds of experiments anymore.)

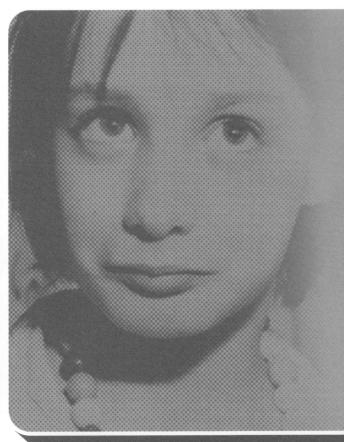

GENIE, THE FERAL CHILD

In 1970, child welfare workers in Los Angeles discovered a child chained to a toilet in a dark room. The father started locking her up when she was only a year and a half old. The girl, Genie, was now almost 14. With almost no stimulation, Genie had never learned how to speak. Child psychologists worked with her relentlessly to teach her English, but while she learned some basic social skills and non-verbal communication, she never fully acquired a first language. This supported the theory that humans have a "critical period" for language acquisition and may explain why it is so much easier to learn a new language when we are young.

LEARNING STYLES DEBUNKER

What kind of learner are you? You may have been asked this question in school or by a friend or maybe even a therapist. Surely you have a preferred learning style. Are you a visual learner? Maybe you retain stuff best when it's auditory information. Some people can pick up a book and learn everything by reading. Or perhaps you're a kinesthetic learner? The concept of learning styles first started to spread in the 1990s and it seemed to explain why teachers struggled to connect with every kid in the classroom, or why certain students had a hard time picking up the material. Thousands of schools conducted tests to help students identify their preferred mode of learning. Every child is special and unique, so it makes sense that their learning style is unique, right? Nope.

Reports of My Learning Style Have Been Greatly Exaggerated

In the time since this concept became popular, researchers have published study after study showing that people aren't really one kind of learner or another. That test you took? Yeah, it's actually just telling you what you enjoy more, not what's actually better for your memory. When students are told their learning style and given strategies on how to study more effectively with their particular learning style, they don't do any better on tests. And it turns out that visual learners are no better at remembering images than auditory learners, while auditory learners are no better at remembering verbal info than visual learners.

It Just Depends

So what's the truth? Well, there is some variability between people's learning styles—you probably have strengths and weaknesses in certain areas. But for the most part, it's situational. Your "learning style" changes depending on the task. For example, you may really want to ice skate. You could read about it and watch videos about it and hear someone's advice, but the best learning will happen once you finally step onto the ice.

BANDURA

For a long time, it was assumed that children learned and developed through direct experience and reinforcement of behaviors. In other words, you know how to use a fork because you have practiced using a fork. But wait. Why do you know how to strum a guitar, even if you've never played before? What makes people more likely to abuse their children if they were abused themselves? How did that toxic ten year old on Xbox learn those swear words?

Where Do We Learn Violence?

These questions fascinated Albert Bandura, who is probably the greatest living psychologist right now and seems like a generally good dude. Back in the 1960s, he was studying aggression and wanted to know how some kids become more violent than others. Was it inherent or was there something else going on? Bandura had a hunch that they actually learned it by watching others. To test this, he came up with the iconic "Bobo doll experiment."

Monkey See, Monkey Do

The experiment was simple. One group of children watched a video of an adult being aggressive with an inflatable clown doll. The adult would punch it, yell at it, throw it in the air, and hit it with a hammer. Another group of children watched a video of an adult playing with the doll in a friendly, non-aggressive way. A final group of children didn't watch a video at all. Afterward, each child was allowed into a room filled with toys, including the Bobo doll. What they found was that kids who didn't watch the video or who saw the non-aggressive video either ignored the toy or played with it gently. But the kids who saw the aggressive video closely copied what they saw the adult doing—including hitting it with a hammer! He was training these kids to be serial killers! But not really. After all, this is a *social* learning theory. If the child was reprimanded for attacking the doll, they stopped. This really goes to show that children learn aggression and other social behaviors by observing and imitating others. So be a good role model. You never know who's watching!

MYTHBUSTER: DO VIDEO GAMES MAKE KIDS VIOLENT?

You may have heard that video games are to blame for kids becoming aggressive, lacking empathy, even engaging in mass shootings. But lucky for you, this claim is OP and needs to be nerfed. It is true that, just as Bandura demonstrated in his study, kids feel more aggressive immediately after playing violent video games. But the aggression disappears as quickly as it arrives. Recent long-term studies have shown that playing thousands of hours of violent video games won't change your aggression levels. Don't worry, you can still play *Stardew Valley* without becoming a cold-blooded killer!

Chapter 5

ELECTRIFYING TALES

When we first started studying the biology of the brain, our tools were pretty limited. Think like . . . ice picks (as in lobotomies). But along came microscopes, and the ability to harness electricity—and things started to get interesting.

Philosophers and naturalists had all kinds of cool (and weird, and sometimes bad) ideas about the brain for thousands of years, but it's only been in the last couple hundred that we've been able to actually dig inside our heads and start figuring out how it all works.

This is also around when science stopped being a thing that rich white dudes just played around with in their absurdly large mansions and started focusing on actually using experimental evidence to back up the claims people were making about the brain and mind.

After that, science got noticeably more scientific (well, most of the time, anyway) and there was a shift away from mysticism, toward animal studies and actual biology. Along the way, some folks had some pretty incredible ideas and did some amazing experiments.

THE ENLIGHTENMENT

In 1543, Andreas Vesalius (whom we first met on page 37) published *On the Workings of the Human Body*, a set of books on human anatomy—based on real human anatomy! That's right, Vesalius broke with tradition and dissected a whole bunch of bodies to figure out what we looked like on the inside. Up until then, it was heavily frowned upon to cut open bodies, as they were considered sacred.

His books, and Copernicus' *On the Revolutions of the Heavenly Spheres* published the same year, are widely considered to have kicked off the Scientific Revolution.

Over the next couple of hundred years, there were big shifts in scientific thinking. Weird theories about the body's humors and sperm in the brain were out, and the scientific method was in, baby!

Scientific societies flourished and scientists (mostly rich white dudes) could talk about their ideas and experiments, and everyone got on board with the idea that the Earth goes around the sun instead of the other way around.

The Age of Enlightenment, and the beginning of modern science, relied heavily on empiricism—which emphasized the use of observation and sensory experience to inform beliefs. It was part of a shift away from the domination of religion as absolute authority, and toward free speech and open thought.

Some things had already been pretty well figured out by this time . . . like the fact that gravity was a thing. But other mysteries, like how the brain works, were still on the list. And thanks to the scientific method scientists were closer than ever to figuring some of it out.

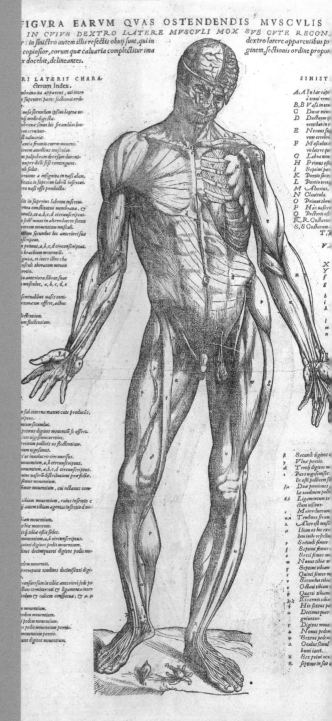

What It Is: Caffeine, found in coffee aka java, bean juice, joe (96 mg per cup), tea (47 mg per cup), energy drinks (29 mg per cup), cola (22 mg per cup), pills (around 200 mg per pill), and powders . . . it's the most widely used psychoactive drug in the world, and it can be found in a wide variety of places.

What Type of Drug It Is: Stimulant

What It Does: Within about an hour of your first cup of joe, and for three to four hours after, caffeine reduces fatigue, enhances wakefulness and reaction time, and improves muscular power and athletic performance. On the downside, the drug can cause insomnia, jitteriness, anxiety, and gut problems. At very high doses, you can experience caffeine intoxication, which can lead to heart palpitations, psychosis, and in rare cases, death.

How It Does This: Caffeine binds to and blocks adenosine A2A receptors in the brain, which leads to a decrease in the levels of chemicals associated with drowsiness.

What the Risks Are: Caffeine really isn't bad for you. Actually, it might be kind of good! You may have been told that it would stunt your growth as a kid, cause cancer, or get you addicted. In reality, it's considered pretty safe. In fact, there's some evidence that caffeine can help prevent Alzheimer's and Parkinson's Disease. So drink up that brain juice! (In moderation, of course.)

SHOCK AND AWE

During that time, there were ongoing arguments about how muscles in the body functioned. Some folks thought that they must be like little balloons, filling up with water or air or "animal spirits" as they contracted. But there was no evidence that muscles expanded the way they would if they were tiny balloons, and Galvani was about to set fire to that theory with one little spark.

The Game Changer

Here's how the story goes: Galvani and his wife, Lucia Galeazzi, were conducting some experiments on static electricity using frog skin—bummer for the frogs, but great for Galvani! As he was skinning one frog, Galvani's assistant accidentally touched the exposed sciatic nerve in the frog's leg with a charged scalpel, discharging the electricity and making the leg jump.

This got the wheels turning in Galvani's head and he hypothesized that a force called "animal

There was no shortage of weird experiments to be found during the early days of neuroscience—mostly because we knew so little about how it all worked, so pretty much all guesses were fair game. Including jabbing electrodes into Kermit.

An Unwilling Scientist

Luigi Galvani was an early example of someone whose parents pushed him to pursue a "more practical" career path—he wanted to study theology, but they convinced him to go into medicine. At the time (in the late 1700s), medical training at the University of Bologna was still focused mostly on outdated texts from Hippocrates and Galen. Eventually, he became a lecturer in anatomy at the university and got interested in medical electricity.

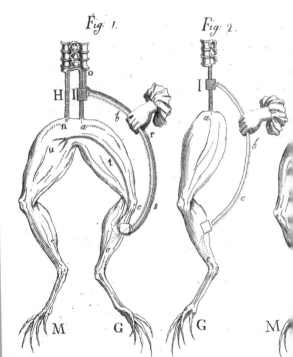

Fig. 1. *Fig. 2.*

electricity," similar to static electricity, flowed through the body through an electric fluid to power muscles for movement. He thought that this force was special, unique to living things.

Galvani was sort of in competition with this other scientist named Alessandro Volta, who wasn't on board with that theory. Volta was pretty sure that it wasn't that there was electricity in the frog's leg, but rather that frog tissue could conduct the electricity between the metal tools used in the dissection. His disagreements with Galvani led him to later invent an early simple battery, using zinc and copper, to prove his point. Nothing like a bit of friendly professional competition to supercharge your productivity!

The Galvanis' electrical experiments kicked off the field of electrophysiology, and had an important role in helping scientists understand that nerves use electricity to send signals, instead of spirits or flowing liquids.

Science Gets Spooky

In a very Frankensteinian twist, Galvani's later experiments included harnessing lightning to make frog legs jump—which might have been the inspiration for his nephew's later work using electricity to "reanimate" the dead. A physician and physicist, Giovanni Aldini was interested in many topics, but he's perhaps most famous for using electrostimulation on the corpse of George Forster, an executed criminal. The application of an electrical charge to the body made its face twitch, fists clench, and even move its legs.

MARY SHELLEY

If all this talk of using electricity to reanimate corpses reminds you of a certain classical science fiction novel, it's no coincidence. When a young, bored Mary Shelley was pulled into a ghost story contest on a rainy day in Switzerland, it turned out that her recent study of Galvani's writings were pretty influential on her work. She mentioned galvanism as an influence in the story. But despite the pop culture depictions of electrical apparatuses and lightning strikes being harnessed to bring the Monster to life, Shelley only briefly mentioned the actual creation of the creature in her novel; there's no telling which side of the animal spirits or animal electricity debate she actually came down on.

GOLGI AND CAJAL GET THE PICTURE

When Golgi and Cajal showed up on the scene, it had only been within the past century that scientists realized the body was made up entirely of microscopic cells, and that these cells all played different roles in the body. Golgi was about to invent a game-changing new technique for looking at individual brain cells that would also kick off a rivalry between the two scientists, and help cement a new understanding of how the brain is built.

Golgi's Theories

With all that brain tissue clumped up together, it was pretty hard to figure out what, exactly, the different cells looked like—or how they worked.

In the late 1800s, biologist Camillo Golgi invented a new way to visualize brain cells, called the "black reaction" (sounds ominous!) that used incredibly toxic chemicals to first harden the tissue and then stain portions of it black. Somehow, this particular technique only stained some brain cells—but it always stained the individual cells in clear detail.

Golgi did a lot of amazing things, like helping to prove that malaria was caused by a parasite, and discovering the Golgi apparatus, a piece of cellular machinery that's critical for packing up and shipping out proteins. But he wasn't right about everything: He was pretty firmly convinced the nervous system was made up of one really long, intricate network of nerve fibers.

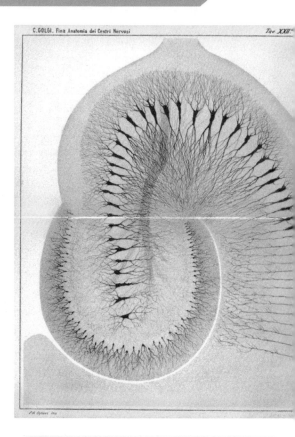

C.GOLGI. Fina Anatomia dei Centri Nervosi   Tav XXII°

THE MYSTERY OF THE BLACK REACTION

Golgi and Cajal were able to generate some of the earliest sketches of individual brain cells thanks to the "black reaction," a technique invented by Golgi to stain brain tissue that uses the incredibly toxic chemical potassium chromate and silver nitrate. When the tissue is cut and placed on a slide, some of the individual neurons are completely stained black, showing their axons and dendrites in all their glory. The wild thing is, 150 years later, we still have no idea why, exactly, it only targets some neurons and not all of them. For whatever reason, this approach seems to only stain neurons, and leaving the tissue in the solution for longer leads to more cells being stained—but still no one knows exactly why some cells get stained first while others don't.

Gifted and Talented

Santiago Ramón y Cajal didn't fit many of the stereotypes of a scientist when he was young: He was a troublemaker, not a nerd, and he wanted to be an artist, not a scientist. But under his father's coaxing, he ended up studying anatomy, and eventually got interested in histology—the study of the microscopic structures in tissue.

He learned about Golgi's method for staining brain tissue, and applied it to his own research. He ended up completely enamored with the delicate, complicated world of neuronal cell bodies, axons, and dendrites that he saw through his microscope, and would spend hours sketching what he could see. His careful observation and attention to detail helped him make some big discoveries.

Ahead of His Time

Cajal is responsible for our understanding of the Neuron Doctrine: the reality that the nervous system is made up of discrete, individual cells called neurons, and that these neurons are connected with one another to send information throughout the body. But since Cajal couldn't actually see the connections between the cells, his debate with Golgi lasted for decades. It wasn't until the 1950s, when the electron microscope was invented, that scientists were

finally able to get up close and personal with the neuronal synapse and prove that neurons were discrete cells.

Cajal was also the first person to note the unique structure of most neurons; he saw that they usually had a cell body surrounded by lots of little projections, called dendrites, and one long projection leading away from the cell body called the axon. Later, scientists would learn that the dendrites receive incoming information while the axon relays those signals out to other cells.

Cajal was so prolific that he published over a thousand of his drawings in a huge, two-volume book set. His classic drawings of the nervous system are still referenced by scientists today, and are not just accurate, they're also beautiful.

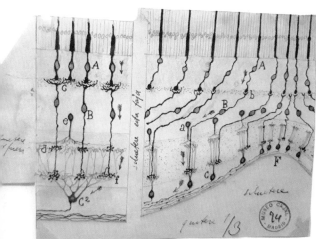

CAJAL'S DRAWINGS

Santiago Ramon y Cajal wanted to be an artist when he grew up, and even though he took the long road there, his dream came true. His prolific work as a scientific illustrator gave us some of our earliest glimpses of what brain cells looked like, and are still considered amazingly accurate and useful, even today.

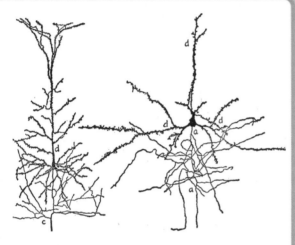

On the left, Cajal shows a pyramidal neuron from the cortex of a rabbit; on the right, a pyramidal neuron from a cat.

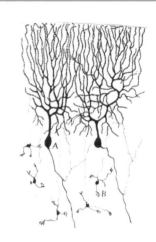

Purkinje (A) and granular (B) cells from the pigeon cerebellum.

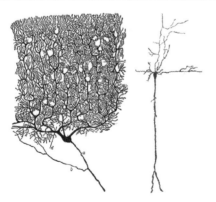

One of Cajal's classics: a Purkinje neuron from the cerebellum of a pigeon.

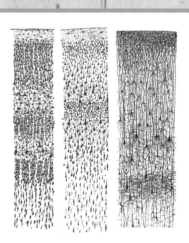

Sections of different regions of the human cortex, using Nissl stain (left and center) and Golgi stain (right).

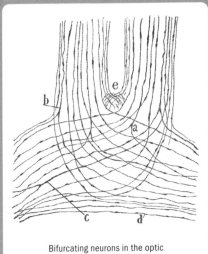

Bifurcating neurons in the optic
chiasm of a rabbit.

Drawing of the optic tectum of a sparrow.

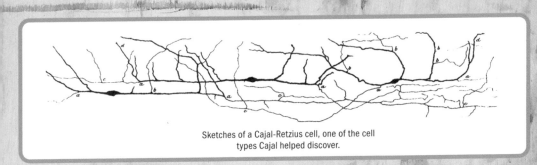

Sketches of a Cajal-Retzius cell, one of the cell
types Cajal helped discover.

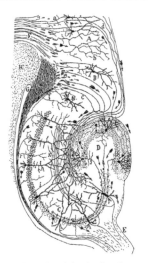

A drawing of the circuitry of
the rodent hippocampus.

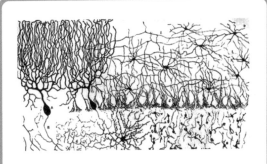

Sketches of the cells in the cerebellum of a chick.

GIANT SQUID AXONS

Calamari on the Menu

You've heard of popular lab animals like mice and frogs. You've maybe even heard of scientists studying birds, or flies. But what about squids? Turns out, scientists have actually learned a lot from squid; more specifically, from the giant squid axon. Now, just to clarify, I mean the "giant" squid axon, not the "giant squid" axon. We're not talking about the kraken here. Nope, just the humble, ordinary squid.

Galvani and his pals had figured out that electricity was a critical part of how the nervous system functioned, but we weren't too sure *how* it all worked. We now know that neurons send signals thanks to an electrical potential across the membrane—basically, there are more negative ions *inside* of the cell than outside. When a signal comes along, those ions all get tossed around, flowing in and out of the cell, passing information along to the next neuron.

It's Electric!

When neuroscientists Alan Hodgkin and Andrew Huxley were first studying action potentials in the mid 1900s, they didn't have the tools necessary to measure these potentials in humans. Instead, they chose to experiment with the longfin inshore, a small squid. Although these creatures are only one to two feet long, the giant axon measures up to 1.5mm in diameter—about the diameter of a pencil lead—making it more than a thousand times wider than typical human axons.

So it was comparatively easy for Hodgkin and Huxley to jab an electrode into this axon. This allowed them to record the first-ever action potential: the spike of a neuron sending a signal. Armed with new tools such as the voltage clamp and chemical inhibitors, Hodgkin and Huxley were further able to pick apart the dynamics of the action potential to help us understand what, exactly, is going on inside a neuron when it's firing.

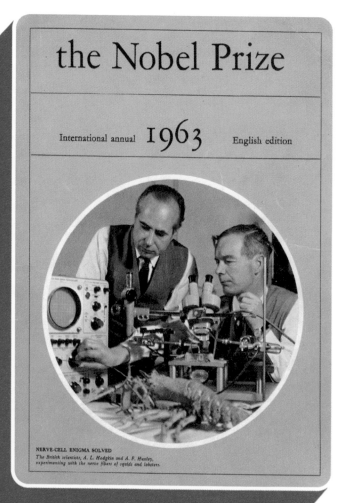

the Nobel Prize

International annual 1963 English edition

NERVE-CELL ENIGMA SOLVED
The British scientists, A. L. Hodgkin and A. F. Huxley, experimenting with the nerve fibers of squids and lobsters.

MY OH MYELINATION

You might be thinking, "Why would a dinky little squid need such a huge axon?" Well, the axon connects to the water jet propulsion system and, in the wild, this allows the squid to rapidly escape from predators. Wide axons transmit electrical signals much faster than thin ones, as the added thickness allows a larger number of electrons to flow through at any given time. If this was an internet cable, it would literally mean more bandwidth!

Humans can get around this limitation because we have myelination—this means our nerves are insulated to preserve conduction. Myelination exists thanks to special glial cells called oligodendrocytes, who put out big arms of fatty cell membranes and wrap them around and around the axons of nearby neurons. These wrapped up membranes act as insulation for the neurons and help transmit signals faster, by holding the ions close to the cell. Instead of myelin, squids just evolved an enormous axon— after all, the faster they could escape, the more likely they were to survive and be able to pass on that trait to their offspring.

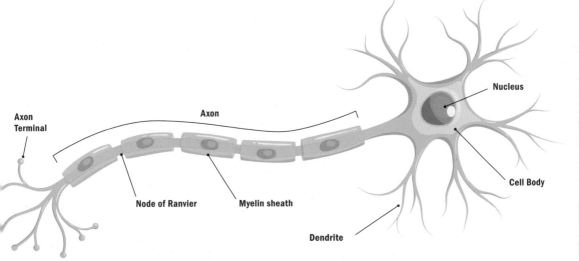

Axon Terminal

Axon

Nucleus

Node of Ranvier

Myelin sheath

Cell Body

Dendrite

THE BRAIN'S SUPPORT CELLS

Turns out, despite all the drama about neurons, they're not the only cells in your brain. About half of your brain is actually made up of a class of cells called "glia"—literally, brain glue. And they were named that because that's pretty much what scientists thought they did for a few hundred years: they hold everything together, like scaffolding for your neurons. Because glial cells aren't electrically excitable, we sort of assumed they didn't really do anything else.

Now, we're pretty biased, because Alie did her dissertation research on glial cells. But believe us when we say that most neuroscientists have vastly underestimated these amazing cells for a long time. Each kind of glial cell is an important part of the whole of the brain; if neurons are the royalty of the brain, glia are her loyal subjects, making sure she's protected and taken care of. Want to learn more about these glorious glia? Check out page 108.

Chapter 6

OH, BEHAVE!

Thanks to the shocking work of psychologists such as Freud and Jung, the study of the human mind had quickly become a pop culture curio. But there were others in the field who found this status off-putting, and took psychology from psuedoscience to real science.

The initial research in psychology was . . . squishy. Although Freud and Jung and other early adopters brought psychology into public awareness, their study of the mind was largely subjective and left much to be desired.

Psychology was not a "science" yet. But boy, oh boy, it was about to go through a monumental shift with the birth of behaviorism. And in short order, the behaviorists became the "cool kids" on the block.

PAVLOV? THE NAME RINGS A BELL

Wanna know why Ivan Pavlov's hair was so soft? Classical Conditioning. (B-dum-tsch!) That joke might make a little more sense in a bit.

Ivan Pavlov was a very good boy. He behaved well, took care of the family, and received high marks in school. His father was a Russian Orthodox priest, and Ivan followed in his dad's footsteps with his own goody-two-shoes by pursuing religious studies. However, partway through the seminary, Pavlov lost his faith and, in pursuit of knowledge, felt called to a different vocation: science!

An Appetite for Knowledge

Pavlov dropped out of priest school and started studying physiology instead. Here, he flourished—he won awards and was praised for his stellar work in the field. Pavlov's hunger for science grew and, after graduating, he became

unusually interested in digestion. In particular, he was fascinated by the saliva reflex of the digestive system. Pavlov noticed that the dogs he worked with started salivating just at the sight of the researchers who fed them. How is that possible? Salivation is automatic! He called this "psychic secretion," which sounds really gross. However, this observation led Pavlov to his most famous experiment yet, which unlocked a new era in psychology.

What You Probably Learned In School

Most likely the story you heard in Psych 101 about Ivan Pavlov's groundbreaking work went something like this: Pavlov knew that salivation was an automatic reflex, and not a learned behavior. Likewise, seeing, smelling, and eating food makes you salivate; that all makes sense. But could you trigger salivation with something totally unrelated? Pavlov tested this out with dogs by ringing a bell right before feeding them. Initially, the dogs didn't respond to this sound— it's just a random bell. But after a few days of ringing it before food time, the dogs started to salivate just at the sound of the bell, even if there was no food. They associated the sound of the bell with the arrival of food, and were trained to salivate to something that Pavlov picked at random. He made them salivate on command! This is called classical conditioning, and it has become the foundation of behaviorism, which believes that all of life's secrets may be revealed through the observation and understanding of behavior.

What They Didn't Teach You in School

First, let's get this out there: Pavlov did not use a bell. I bet your college professor put it on your final exam. It. Is. Wrong. So what did he use instead? Funny enough, a lot of other things like buzzers, whistles, and electric shocks (yikes!).

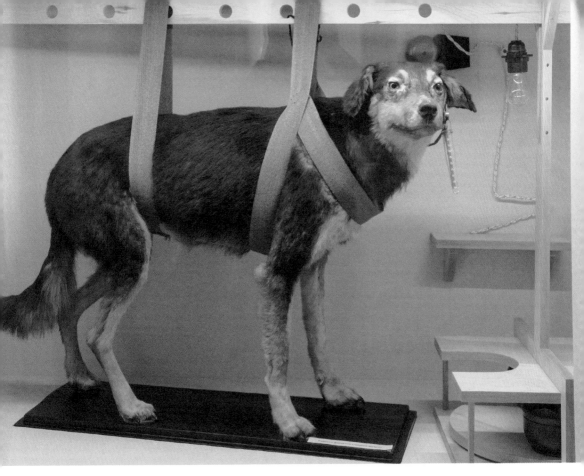

But the OG stimulus was a metronome. Pavlov trained those dogs so well that not only did they salivate to the sound of the metronome, but he could train them to salivate to a particular speed.

Also, Pavlov tends to get lumped in with other behaviorists, like the ones we're about to talk about. However, the reality is that Pavlov was not a behaviorist himself. He believed that the human brain was kind of a "black box," never to be fully understood.

WORTHY OF NOTE:
HE WAS NOT NICE TO DOGS

Pavlov wanted to repeat his experiments frequently, with as few dogs as possible. But it's hard to study digestion if the dog isn't hungry. To speed things up, Pavlov removed each dog's esophagus and made an opening in the throat so the food would fall out of the hole as the dog ate, never making it to the stomach. He also connected the dogs' salivary glands to tubes to collect the saliva, which he bottled and sold as a remedy for dyspepsia. Unfortunately, without these essential fluids, most of Pavlov's dogs died within a week. Poor things.

DR. JOHN B. WATSON

John Broadus Watson also grew up in a very strict, religious household. His mother, Emma Watson (not *that* Emma Watson), ran her household like a scene out of the movie *Footloose*: no drinking, no smoking, and certainly no dancing! She hoped to raise her young boy to become a minister, but John's upbringing left a sour taste in his mouth. Like Kevin Bacon, he felt restricted and spent his evenings dancing angrily to Kenny Loggins . . . Okay, that last part isn't true. But apparently he was a terrible student, had no friends, got arrested twice and, in direct defiance of his mother, converted to atheism. He had no problem going against the grain, which became quite apparent in his career as a psychologist.

Get Out of My Head!

Although he was a part of the field, Watson grew displeased with psychology's trajectory and proposed a pretty radical idea: Stop studying this introspective mind crap because the mind doesn't exist. He told his fellow psychologists, "never use the terms consciousness, mental states, mind, content, introspectively verifiable, imagery, and the like." Madness! Psychoanalysts will be out of a job! Psychology is the analysis of the psyche. If you can't study internal thoughts, then what's left?!?

His answer? Behavior! Since you can't see a person's thoughts, Watson suggested that psychology adopt a more rigorous scientific approach and focus only on what you can directly observe. This idea took the psychology world by storm and we've felt its ripples ever since.

What Watson Got Right

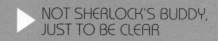

Watson was correct that psychology needed hard evidence to be more in line with the biological sciences. And we can learn a heck of a lot from behavior. He deserves credit for demanding that experimental psychology use the scientific method. His staunch views about the impact of one's environment means he was also very vocal against the eugenics movement, so that's good.

What Watson Got Wrong

Watson wrongly exaggerated the importance of "nurture" over "nature," and thought he could control most everything about a person through their environment. I mean, just look at his quote below (although it was a bit tongue-in-cheek). He also suggested that parents treat their children like little adults by shaking their hands instead of kissing or hugging them to prevent kids from becoming "soft." And you know the "cry it out" method for babies to learn how to sleep independently? Watson promoted that, too, but research has shown this can have pretty detrimental effects on development. To his credit, Watson later said he "didn't know enough" about children and admitted fault.

"Give me a dozen healthy infants, well-formed, and my own specified world to bring them up in and I'll guarantee to take any one at random and train him to become any type of specialist I might select—doctor, lawyer, artist, merchant-chief and, yes, even beggar-man and thief, regardless of his talents, penchants, tendencies, abilities, vocations, and race of his ancestors."

THE LITTLE ALBERT STORY: THAT'S NO ORDINARY RODENT!

Watson wasn't always such a good guy. He had an affair with a student half his age, was super misogynistic, and conducted some really unethical studies. In one experiment, Watson wanted to see if he could instill a fear (or phobia) in humans. To do this, he placed a nine-month-old baby named "Albert" on the floor and encouraged him to play with a fuzzy white lab rat. As soon as curious Albert tried to touch the rat, Watson would bang a cymbal with a hammer, scaring the poor kid half to death. After doing this over and over, Watson once again put the rat in front of Little Albert and he cried . . . without ever hearing the cymbal. They did such a good job terrifying this young child, he became afraid of anything remotely furry. Watson never got around to deconditioning Little Albert's fear, and we may never know what long-term effects this study had on his life.

NEURO TRANSMISSIONS FILM CORNER:
BRAINWASHING

Around the 1950s, "brainwashing" became a terrifying prospect. Governments were researching how to rewrite people's minds to bend their will. It quickly captured the minds of the public and, although it still lacks scientific evidence, brainwashing was great for Hollywood!

Captain America: The Winter Soldier (2014)

S.H.I.E.L.D. gets hijacked by Hydra baddies who have a secret weapon: a mind-controlled supersoldier (and Captain America's oldest friend)—Bucky Barnes.

Depiction of Brainwashing: ★★☆☆☆
The idea that Bucky would unwillingly become a mindless killing machine after hearing a string of Russian words is laughable.

Plot: ★★☆☆☆
I love superhero movies, but I found this one frenetic and a bit shallow on how they dealt with morality.

A Clockwork Orange (1971)

In dystopian Britain, delinquent criminal Alex is "rehabilitated" by strapping him to a chair and clamping his eyes open to force him to watch violent films until he becomes sick.

Depiction of Brainwashing: ★★★★☆
Although the film's visuals are extreme, aversion therapy is a real thing and can be highly effective.

Plot: ★★★★★ It's dark and violent, yet Alex is witty and childlike. The combo makes it disturbingly enjoyable.

Full Metal Jacket (1987)

After weeks of harsh punishment, ridicule, and hazing, Leonard Lawrence transforms from an incompetent recruit into a deadly marksman. But Leonard seems . . . changed.

Depiction of Brainwashing: ★★★★★
This accurate portrayal of brainwashing shows a shift in Leonard's persona and ensuing mental breakdown as a result of extreme trauma. That's the real stuff.

Plot: ★★★★☆ The first half of the movie is the best part, but it's still a fantastic film about the horrors of war.

The Manchurian Candidate (1962)

A former prisoner of war is brainwashed by Communists and becomes an unwitting political assassin.

Depiction of Brainwashing: ★★☆☆☆ Brainwashing doesn't remove a person's free will or make them into sleeper agents. And seeing a particular playing card certainly wouldn't trigger a hypnotic state.

Plot: ★★★★☆ Despite being a Red Scare film, it's a classic psychological thriller with a great plot.

Zoolander (2001)

Male model Derek Zoolander is brainwashed to kill the prime minister of Malaysia to the song "Relax" so that evil fashion mogul Mugatu can keep operating child labor factories.

Depiction of Brainwashing: ★☆☆☆☆ Derek is apparently easily brainwashed because he's super foolish, but there's no actual evidence to support this idea. But listen, it was never meant to be accurate.

Plot: ★★★☆☆ It's a deliberately goofy comedy about male models that genuinely makes us laugh out loud.

REAL LIFE CASES

When we think of brainwashing, most likely we think of unwilling mind control. But evidence to date indicates that it's simply not possible. Nowadays, the term "brainwashing" acts as shorthand to describe cultish devotion, coercion, and uncharacteristic behavior. Let's look at some real-life examples, shall we?

Project MKUltra (1953-1973)

The CIA tried real hard to mind-control people using hallucinogens, hypnosis, and just plain torture. This was very, very illegal. The US government was looking for the perfect "truth serum," and pretty much everyone involved in the project was unwittingly drugged with LSD at one point or another. As far as we know, MKUltra never successfully brainwashed anyone.

Jonestown (1954-1978)

This is where the term "drinking the Kool-Aid" comes from. However, that's a bit of a misnomer. In 1978, over 900 members of the Peoples Temple died after drinking cyanide-laced Flavor-Aid. The cult leader Jim Jones had brainwashed his members to follow his fatal orders, while others were forced to do so at gunpoint.

The Manson Family (1968-1975)

On an abandoned film plot in the desert, a commune of drugged-up hippie girls were ordered by their leader, Charles Manson, to go on a killing spree of famous actors and socialites to hasten the apocalyptic race war. The news said they were brainwashed, but most likely these girls did it of their own volition.

NXIVM (1998-2018)

This multilevel marketing scheme attracted wealthy women who were seeking personal betterment. Members literally bought into the company's bizarre beliefs and practices. But you may know it best as the cult that got *Smallville* actor Allison Mack arrested for sex trafficking.

WHAT ABOUT STOCKHOLM SYNDROME?

In 1974, Patty Hearst was kidnapped from her college apartment by a terrorist organization. Two months later, she enthusiastically wielded a gun while robbing a bank with her captives. What the heck happened? Stockholm Syndrome supposedly stems from human survival instincts, though it is not a recognized condition in psychology and some call it a myth invented to discredit female victims who did not trust authorities to help. The FBI say only 8 percent of hostages start sympathizing with their captives, form close emotional bonds, or even join forces with them.

SAN MATEO SHERIFF
HEARST   P C
9 19 75   106234

B.F. SKINNER THINKS INSIDE THE BOX

Many American psychologists didn't think much of Freud and company's psychoanalytic approach. By taking a more scientific approach, these so-called behaviorists moved psychology from the couch to the laboratory.

Humble Beginnings

In a small railroad town, a young man named Burrhus Frederic Skinner sat in his parents' basement with a blank page in front of him—writer's block. He had just graduated with a literature degree and he hoped to write a masterpiece, but it simply would not come. A whole year passed and Skinner (who went by B.F. for obvious reasons) needed a change of pace. He moved to New York and took a job working at a bookstore. It was here that B.F. encountered books by two folks you may have heard of: Ivan Pavlov and John B. Watson. Eureka! Impressed by what he read, Skinner gave up on being a great writer and decided to pursue psychology instead.

A Chip on His Shoulder

Skinner was a pretty pompous dude. He often felt smarter than his Harvard peers, and viewed his study of behavior as superior to those trying to understand the mind. Annoyingly, he actually was pretty smart and realized that Pavlov's classical conditioning didn't explain all behavior. See, Pavlov only studied reflexive behaviors like salivating. But how do we pick up on other things like learning to play golf or learning to not touch a hot stove? Skinner thought that these kinds of behaviors are learned through consequences. This led to his most important contribution to science: operant conditioning.

Operant conditioning has a basic premise. If you reward a behavior, that action will be strengthened and be more likely to happen again. If you punish a behavior, it will weaken and decrease the likelihood of it happening again. So simple, yet so perfect. Skinner's controlled experiments on various animals supported this concept; Skinner himself believed this meant there was no such thing as free will!

What's In The Box?

In order to best study behavior in animals without any other distractions at all, Skinner invented his most famous apparatus: the Skinner Box. In its simplest form, it's a box with a lever and a food dispenser. Skinner put a rat (or some other animal) in the box and waited. The rat explored the box and eventually hit the lever, triggering a food pellet to drop. When the rat hit the lever again, it once more dispensed a pellet. After several repetitions, the rat learned that touching the lever equals food.

But the box can get more complex: Skinner added lights and an electric grid on the floor. If the rat pressed the lever when the light was green, they got a food pellet. When the light was red, they received a mild electric shock. Needless to say, the rat quickly learned to press the lever only on the green light. But be careful! Sometimes the animals would get too good at it and repeatedly pressed the lever to gorge themselves.

What Skinner Got Right

Skinner exposed something truly crucial about learning: A big chunk of it happens through reinforcement and punishment of behaviors. We associate those consequences with what we're doing, and it shapes what we do next. Strangely, he even developed pigeon-guided missiles for World War II, by putting pigeons in the nose cones of missiles and training them to peck at targets. It was effective, but never got off the ground because nobody took him seriously.

What Skinner Got Wrong

Skinner's theory doesn't account for everything. Contrary to operant conditioning, commenters on social media who received more negative votes ended up commenting more often. Why? And hey, what about human speech? It's one thing to give consequences for accidental behaviors, but what about novel verbal commands? Skinner didn't seem too interested in language—except when it came to pigeons (he really loved those birds). He thought he'd taught pigeons to read all sorts of words, but it turns out that they're just really good at pattern recognition!

BEHAVIORISM IN THE WILD

Some behaviorists became pop-culture figures and their ideas reached far beyond academic psychology. Intrigued by these newfangled concepts, some devotees took behaviorism to . . . utopian lengths.

In 1949, Skinner achieved his lifelong dream of being an accomplished author and published *Walden Two*. This "novel of ideas" was about an experimental community in which the open-minded inhabitants experimented with behavior modification, scientific child-rearing, equal rights for women, and other frightfully modern notions. While the book was considered controversial at the time for its rejection of the spiritual realm and associated concepts—like, ya know, free will—it went on to inspire the establishment of over a dozen real-life communities, some of which are still in existence today. The postcards are our invention, but the features touted are all too real.

Greetings from

The Intentional Community of

TWIN OAKS, VIRGINIA

HARD WORK * SHARED INCOME * NO TV!

Feeling Trapped?
"That's Normal" Says Our Founder!

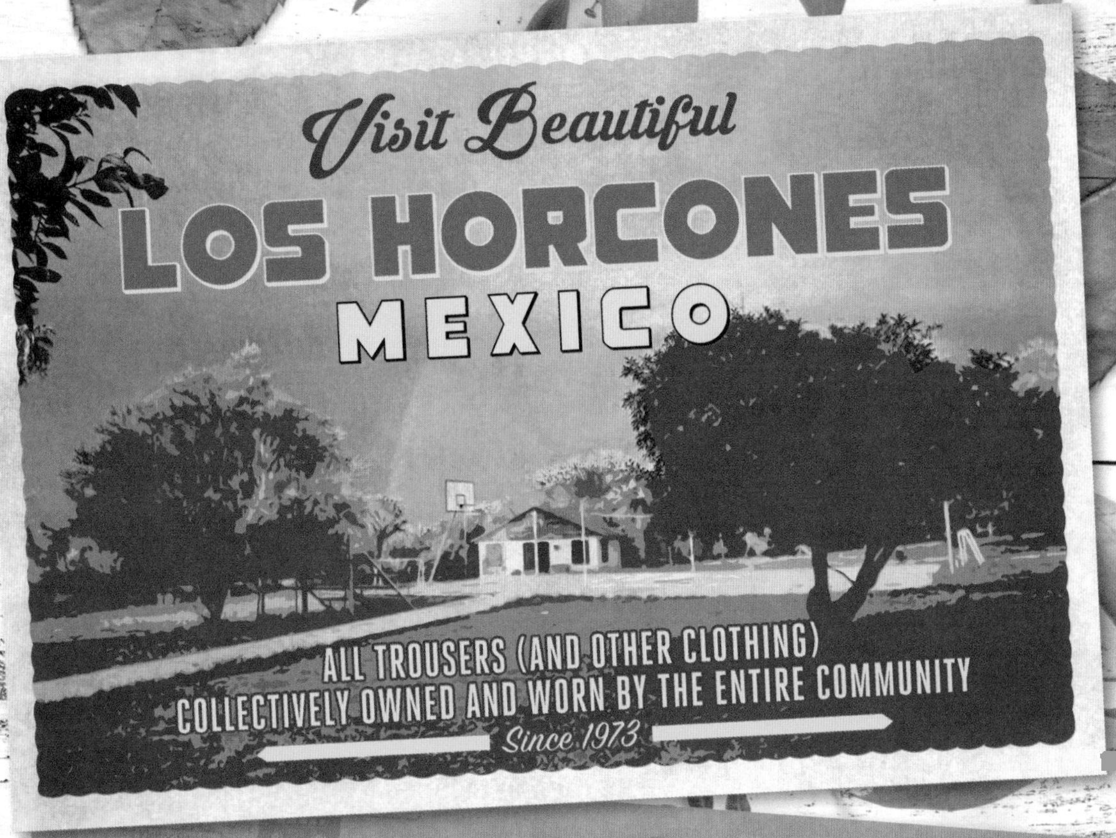

Visit Beautiful
LOS HORCONES
MEXICO
ALL TROUSERS (AND OTHER CLOTHING)
COLLECTIVELY OWNED AND WORN BY THE ENTIRE COMMUNITY
Since 1973

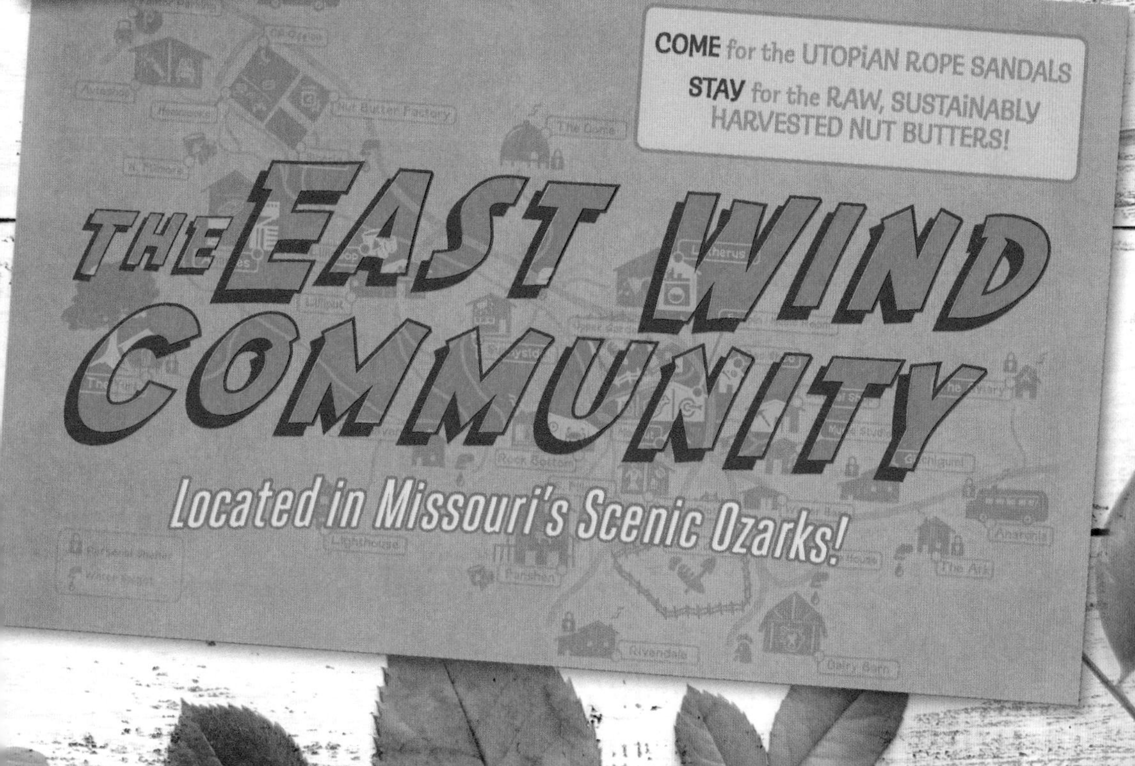

COME for the UTOPIAN ROPE SANDALS
STAY for the RAW, SUSTAINABLY HARVESTED NUT BUTTERS!
THE EAST WIND COMMUNITY
Located in Missouri's Scenic Ozarks!
Nut Butter Factory
The Dome
Rock Bottom
Lighthouse
Rivendale
Dairy Barn
The Ark

COGNITIVE THEORY

Behaviorism was all the rage in psychology for over thirty years. It was psychology you could see. Who cares about what's going on inside your head? Well, cognitive theorists cared. And they pushed back!

By the 1950s, behaviorists had gained notoriety for their experimental methods. They had changed the field forever, and psychology was on track to predominantly become the study of behavior. But this methodology neglected to address some pretty big questions. Like, how do you explain and understand unobservable processes? How does behavior justify thoughts? Imagination? Ideas? Y'know . . . consciousness?

Skinner would scoff at these questions, since he did not believe consciousness existed. Other less-radical behaviorists tried to explain internal processes by saying that maybe they were a response to stimuli from the outside world or were stimuli themselves that provoked behavior. Clever!

Mister Roboto

Some folks were dissatisfied with this answer. World War II brought about new technological developments, including computers that could solve complex logic problems. If a computer could work out these problems without being "trained" or coded to answer that specific question, then couldn't humans be the same? This introduced a new way of understanding consciousness, which viewed the brain as an information processor. Like a computer, we filter input information, interpret it, and then either act on it or store it in memory.

Skinned Alive

As this idea spread, other skeptics spoke out against behaviorists who they believed had gotten fat on an incomplete theory. The leader of this charge was linguist Noam Chomsky. B.F. Skinner had recently published a book proposing that behaviorism can explain how humans acquire language. Chomsky systematically and savagely tore apart Skinner's argument by showing that it could not account for how much language children learn in such a short time. Chomsky contended that we are simply not exposed to enough language in the natural environment while growing up to learn the infinite number of complex variations of syntax and grammar.

In its place, Chomsky proposed his own theory that humans have an innate cognitive process that allows us to organize and expand on the language that we're exposed to. In other words, we're not programmable automatons who simply spit out what's fed into us. Instead, we're more like machine-learning AIs that learn and improve from experience, without being explicitly programmed. That means humans possess the complex brain wiring and mental processing power to do all of this from birth. This theory flew in the face of behaviorism. Chomsky suggested that language learning was both biological *and* cognitive!

These revelations started the "cognitive revolution," which gave birth to cognitive science, an interdisciplinary field that retains the same scientific rigor of behaviorism. But this time, instead of studying behavior, it studies the mind and its processes, including perception, attention, memory, reasoning, and language. *¡Viva la revolución!*

NOAM CHOMSKY & THEORIES OF LANGUAGE

Famous linguist Noam Chomsky did not buy into the age-old beliefs of philosophers and behaviorists that we come into existence as a "blank slate." Instead, he suggested that every human is born with the ability to understand and use language. We have an innate capacity to sort words into different categories (like verbs, nouns, and adjectives), allowing us to combine these categories in new, meaningful, grammatically correct ways. To put it plainly, every human has a Universal Grammar, regardless of what dialect you speak. You just have to learn the vocabulary!

SERIOUSLY BAD SCIENCE

Of the handful of modern psychological studies most commonly held up as needlessly cruel, procedurally flawed, or both, we've already discussed the Little Albert study earlier in this chapter. If you'd been wondering about the other four points in that pentagram, you're in luck! . . . unlike the range of innocent animals, children, and grad students in these studies.

THE MONSTER STUDY:
BECAUSE ORPHANS DON'T HAVE RIGHTS, RIGHT? (1939)

The Study: Twenty-two orphans were divided into two groups. Half received positive speech therapy where they were praised for speaking so well. The other half received negative speech therapy, were criticized for speech imperfections, and were told they were developing a stutter.

What It Was Supposed to Tell Us: Could you induce stuttering in healthy children through negative feedback and could you reduce speech problems in stuttering children through positive feedback?

What We Actually Learned: Being constantly belittled for how you speak will really mess you up! The children who received negative speech therapy showed signs of distress, withdrew at school, and stopped talking altogether. In 2007, seven of the orphans were awarded $1.2 million dollars for lifelong psychological and emotional damage caused by the study.

THE MILGRAM EXPERIMENT:
BECAUSE I SAID SO (1961)

The Study: In different rooms, a "teacher" (the participant) read pairs of words to a "learner" (who was in on the whole thing). When the learner failed to remember the pairs of words, the teacher administered an electric shock that increased by 15 volts for every error. As the shocks got higher, the learner started to yell, then complained about heart issues, then stopped responding altogether. If the teacher hesitated, the researcher in the white lab coat would say, "the experiment requires that you continue."

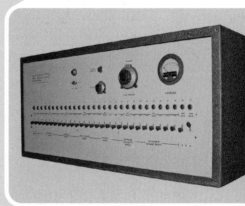

What It Was Supposed to Tell Us: How far would people go in obeying instructions if it involved hurting another person?

What We Actually Learned: It turns out that good people will do terrible things that go against their conscience because an authority figure tells them to obey. Sixty-five percent of participants administered the maximum shock level of 450 volts, despite having no response from the learner.

LEARNED HELPLESSNESS:
THERE'S NO ESCAPE (MID 1960s)

The Study: Dogs were placed in a crate and given repeated mild shocks. Half of the dogs could end the shocks by pressing a lever, and the other half had no way of controlling them. Later, these same dogs were placed in a box with an electrified floor and a low wall that the dogs could easily jump over. Strangely, the dogs who previously had no control of their punishment would simply give up, lay down, and get shocked over and over again, despite having an easy exit.

Shock

No shock

What It Was Supposed to Tell Us: How do animals react to punishment (electric shocks) if they have been subjected to either escapable or inescapable punishment in the past?

What We Actually Learned: Taking away a person's control takes away their hope for the future. Regrettably, the CIA used the learned helplessness model to torture terror suspects after 9/11.

THE STANFORD PRISON EXPERIMENT:
THIS JOB HAS CHANGED YOU, MAN (1971)

The Study: Volunteers were randomly assigned to the role of inmate or guard, and were placed in a mock prison in a basement. Prisoners were dehumanized by calling them by ID number. Guards were deindividualized by giving them uniforms and sunglasses. Dr. Philip Zimbardo, the lead researcher, told the guards to do whatever was necessary to keep order. The guards became so authoritarian and abusive that the study was stopped after only six days.

What It Was Supposed to Tell Us: Is the brutality of prison guards due to their personality types or due to the environment in which guards and inmates are placed?

What We Actually Learned: Ugh, this study was so unscientific that it doesn't really tell us anything. All it really shows is that the participants were behaving in the way they thought Zimbardo wanted.

STUDIES IN ABSURDITY

The dark side of psychological studies and unethical experiments can be very dark indeed. As a somewhat peculiar palate cleanser, please enjoy this real-life research from the stranger side of the lab.

Urine Good Company

Ever get performance anxiety? How about in the restroom? Researchers studied whether having someone at the urinal next to you increases the amount of time it takes to start peeing. Turns out the answer is, "yes!" Having a next door neighbor increases stress, which inhibits relaxation of the urethral sphincter. Even stranger, in order to study this, apparently there was somebody stationed in a stall nearby with a periscope to, ya know . . . see the pee. Wow, so not only an invasion of space, but an invasion of privacy!

Farting to Cope with Existential Dread (as ya do)

This psychoanalytic case study described an abandoned boy who had gone through horrible conditions growing up who had a farting problem. In Freudian fashion, this psychoanalyst concluded that his flatulence was a defense mechanism to "envelop himself in a protective cloud of familiarity against the dread of falling apart, and to hold his personality together."

What's Your Angle?

A group of researchers spent hundreds of hours to discover that physically leaning a particular direction can influence your estimates of things. In particular, they determined that leaning to the left makes the Eiffel Tower seem smaller.

Yeah, I Think It's a Banksy

It turns out that pigeons are extremely good at pattern recognition. How good? Psychologists successfully trained pigeons to discriminate between paintings by Picasso or Monet, even if they'd never seen them before. This study would make B.F. Skinner so proud!

Beauty Is in the Eye of the Beer Holder

They call alcohol "liquid courage," which has now been confirmed by science. A study surveyed intoxicated adults and found that, as they drank more alcohol, they rated themselves as more attractive. That's not too surprising—but they went one step further. Some participants were given non-alcoholic drinks, but thought they'd received the real deal. Surprisingly, people who thought they were drunk (but weren't) also rated themselves as more attractive (they weren't).

Chickens: They're Just Like Us (in that They Prefer Beautiful Humans)

This study raises a lot of questions while answering others. Apparently chickens can tell you who's hot and who's not. Researchers trained some chickens to react to an "average" human face of a particular gender and then showed them a bunch of different faces that ranged from more feminine to more masculine. Consistently, the chickens reacted most to the same faces that human participants had also rated as being most attractive. We're not sure exactly why, but we think it has something to do with our attraction to symmetry.

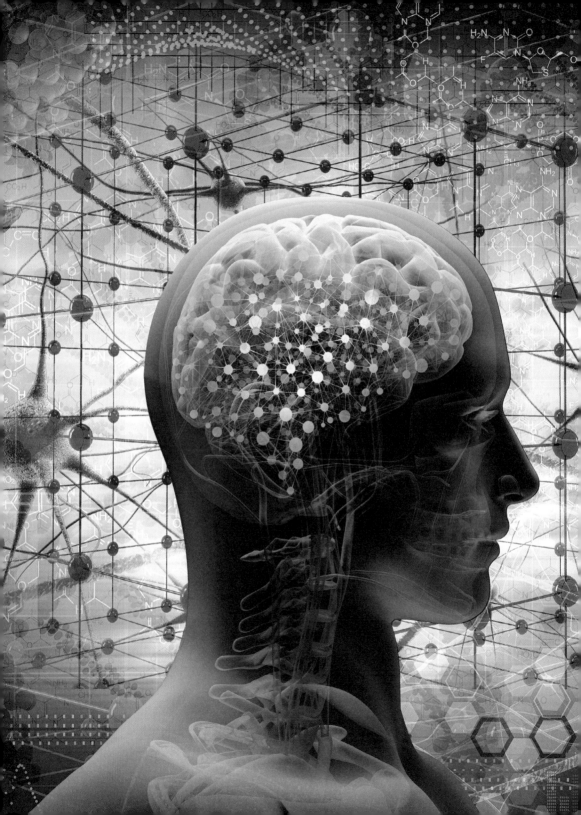

Chapter 7

MODERN ERA, HERE WE COME!

All right, all right, we're almost done with the history lessons, we promise. We've covered everyone from Aristotle to Zimbardo, and how they've all had a hand in shaping our modern understanding of the brain—sometimes in some pretty ugly ways.

But you didn't think they were the only movers and shakers out there, did you? Heck no. Let us introduce you to some of the hottest new brain scientists and the ways they're changing up the field.

In the last century, a lot has changed: We went from Prohibition and the invention of the television to Burning Man and the entire internet in our pockets. Along the way, science has changed a lot too. We figured out the molecular structure of life, in the form of DNA, and used that information to sequence the entire human genome. We created machines that let us look inside the minds of living human brains. And, we started letting women do science!

Haha! No, but seriously. While women and people of color have been active participants in the scientific process for millenia, it's been quite rare for minorities to have a seat at the table—something that's still an issue today. And, it turns out that as much as we'd like to believe that the scientific process is objective, your personal interests, goals, and training have a direct influence on the kinds of questions you ask and how you go about answering them. So having diverse perspectives in neuroscience and psychology means that more diverse questions get asked, more diverse perspectives are considered, and more diverse solutions are developed.

So, as scientists from all walks of life pour into the field and as our ability to pick apart the brain grows, we've discovered some pretty amazing things about how it all works.

INSOO KIM BERG

To start off, let's dive into the life of a local legend in our hometown of Milwaukee, Wisconsin. For most of her early life, Insoo Kim Berg grew up in Seoul, Korea. As was traditional of most Korean families at the time, she had her career decided for her by her parents. Since her family was in the pharmaceutical manufacturing business, she was to become a pharmacist. But Insoo felt stifled by her family and, in the middle of her undergraduate coursework, she chose to move away to the United States with her then-husband.

At age 23, Insoo started her new life in Milwaukee (imagine venturing far from home and learning a new language at that age). She worked with rats all day as a laboratory technician in order to build experience before applying to schools and fulfilling her parents' wishes of becoming a pharmacist. But as she spoke with

her peers, she noticed that they thought of their lives differently. In her own words, "Students in the US had a choice in their area of studies. I was absolutely shocked by that. The idea just blew me away. And so then I got this idea: my parents are seven to eight thousand miles away. They have no idea what I'm doing here. So maybe I could do the same."

Insoo decided that she was done dealing with rats and instead wanted to directly help people. So she switched majors and studied social work at the University of Wisconsin-Milwaukee. To her surprise, her family didn't freak out when she told them about it! There, she achieved both her undergraduate and graduate degrees and found her calling as a clinician. Insoo then traveled the country, conducting post-graduate studies in Chicago, Topeka, and Palo Alto.

It was in California that she met her future husband, Steve de Shazer, who immediately recognized her clinical talent and became enraptured by her. Steve realized Insoo was using novel, effective techniques in her therapy sessions and became interested in documenting this, perhaps to better understand what worked. Insoo was the savant, practicing her craft with ease; Steve was the academic, accurately identifying what she did in therapy. The two of them pored over thousands of hours of recorded therapy sessions to identify what other specific things therapists do that worked—and also what didn't. Combining Insoo's natural abilities and their painstaking research, they developed and proliferated a new kind of therapy: Solution-Focused Brief Therapy.

This technique likely would have flown under the radar if it were not for Insoo Kim Berg's radiant personality and absurdly dedicated work ethic. Others described her as a tenacious, generous, warm woman with a twinkle in her eye. She was an extremely early riser who remained perpetually on the move, writing articles, leading training, and running her clinic. She was clearly passionate about what she did, and it permeated her life. However, that didn't detract from her deep attentiveness to others and gentle optimism. In 2007, in classic Insoo form, she passed away after working out at the gym. She simply went into the steam room to relax and was later found looking as though she was peacefully sleeping. What a life!

SOLUTION-FOCUSED BRIEF THERAPY

Imagine that you are walking into your first session with a new therapist. Your life is a mess and you can feel the weight of these difficulties on your shoulders. This is how it's always been. As you sit on the comfy couch, you start to recount all of your problems from birth to present day. But something strange happens—the therapist stops you and instead asks, "Can you tell me about when you do not experience these problems?"

This is the revolutionary premise behind Solution-Focused Brief Therapy. It is unnecessary to rehash old struggles. The therapist doesn't need to know what happened in a person's childhood. This person came to therapy to get help with their current problem and they probably have a pretty good idea of how to make their life better. So instead of staying stuck in the past, the therapist guides the client to focus on how they would like their life to be.

Solution-Focused Brief Therapy lives up to its name, too. It's intended to help people develop solutions for their issues as quickly as possible so they don't have to continue suffering. That means that typically only five sessions are required on average and can be as few as just one! Despite its low session count, Solution-Focused Brief Therapy works and has been classified as an evidence-based therapy. It seems to be as effective as Cognitive-Behavioral Therapy and Interpersonal Psychotherapy in less time. So if you're in a rush to get help, try it out!

BEN BARRES & THE GLIAL CELLS

Ben Barres is a perfect example of a scientist whose own unique experiences had a dramatic impact on his research, and whose stellar perspectives and mentorship will have a ripple effect on the field for generations to come.

Ben was assigned female at birth in New Jersey in the 1950's. As a kid, he was precocious, obsessed with science, and was described as a "tomboy." As an undergraduate at MIT in the 1970's, he fell in love with neurobiology, despite experiencing blatant sexism at the hands of some of his teachers. He went on to get his medical degree at Dartmouth University, and later turned down a neurologist job to go back to school again for his PhD at Harvard.

During his training, Ben started to notice that while the neurons in the brain got all of the attention from scientists (hence, you know, the name neuroscience), at least half of the

brain's cells weren't neurons. Those billions of cells, collectively known as glia, were relatively mysterious; little was known about their function, but it was clear that they changed radically in the face of injury and disease. He was especially interested in understanding a subtype of glial cell called astrocytes, and whether or not they were helpful or harmful to the diseased brain.

It was around this time that he also started to really struggle with his gender identity. At the time, while he identified as male, he didn't have any idea how to navigate his feelings about his gender. So he chalked his depression up to low self-esteem and continued his training.

It wasn't until Ben was in his early 40s and had landed his tenure-track faculty position at Stanford University studying glial cells and their role in the brain that he first heard the word "transgender" and a lightbulb switched on. With

the support of his closest mentors, he came out to his colleagues and trainees in a letter where he wrote, "I'm still going to wear jeans and tee shirts and pretty much be the same person I always have been—it's just that I'm going to be a lot happier."

Ultimately, Ben felt that being trans did not negatively impact his career, but it did help him appreciate what other scientists had not. One of his major contributions was the development of new methods to isolate different kinds of brain cells in a dish—neurons, astrocytes, oligodendrocytes, and microglia. By isolating these cells independently of one another, researchers could then study each type of cell closely, and learn more about the ways in which different cells interact with one another. With these techniques in hand, glial biologists have begun to unravel the relationship between neurons and glia, and we've started to understand that this "brain glue" is critically important for supporting normal neuronal growth and development. Without glia, our neurons wouldn't be able to grow properly, form the right connections with one another, or maintain their signaling throughout our lifetimes.

Outside of the lab, Ben used his privilege as a tenured professor to advocate for better diversity and support in STEM. He was known to interrupt his own scientific talks with discussions on the importance of preventing sexual harassment in STEM and providing resources for new parents juggling research careers. He even famously started a beef with the then-president of Harvard University who claimed that women were less capable at STEM careers than men!

He walked the walk, too, mentoring dozens of students during his time as a professor, many of whom have gone on to their own prestigious research careers further exploring the roles of glia in the brain. He was so dedicated to his trainees that, when it became apparent that he was losing his fight against pancreatic cancer, he spent his last months updating his letters of recommendation for his students.

STUDENTS INSPIRED BY BARRES

Beth Stevens: Her studies of the brain's immune cells, microglia, revealed that they play an integral role in brain development by actively pruning back unnecessary neuronal connections. She now studies how microglia respond to signals from the complement cascade, an important component of the immune system, to determine which connections to prune—and how that signaling can go haywire in degenerative diseases like Alzheimer's.

Dr. Nicola Allen: She determined that astrocytes secrete glypicans, proteins that are important for helping neurons form connections with one another. Now she studies the proteins produced by astrocytes in order to understand how these cells function, and how that affects neurons and their connections during development, aging, and disease. She was also Alie's PhD advisor! While in her lab, Alie studied how astrocyte function changes in genetic neurodevelopmental disorders and the effects on neuron growth.

Dr. Shane Liddelow: He studies reactive astrocytes—meaning astrocytes whose behavior and function has changed because of inflammation. His work has found that these normally-supportive cells can turn into killers in response to inflammatory signals from microglia, and he is currently researching how these killer astrocytes might be causing some of the damage seen in neurodegenerative disorders like Alzheimer's disease.

GLIAL CELLS

Think of those glorious glia as the Beyonce of the brain—the stars who are running the show—then you can think of glial cells as her entourage: her support crew that makes sure she's looking and sounding fabulous while she's singing her heart out.

Astrocytes These starry little cells are Alie's favorite. Did you know that when researchers looked at Einstein's brain under a microscope, the only difference they could find was that Einstein had a lot more astrocytes than the average person? So there must be something pretty special about them. These cells shuttle nutrients from the brain's blood vessels to the neurons, form scars and help with repair after brain injuries, and help keep the environment in the brain balanced and friendly for neurons. But they also act sort of like Beyonce's sound techs, wrapping thin tendrils around the connections between neurons to monitor and give feedback on their signaling.

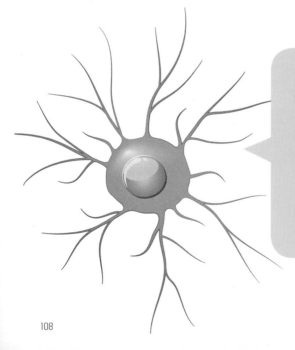

Microglia Basically the brain's immune cells, microglia are like security guards, monitoring the brain's environment and making sure that there are no intruders like viruses or bacteria that might cause problems. When they do encounter a disease or injury, they activate, jumping into action by multiplying and recruiting other immune cells to attack the problem. They also clean up debris and prune back unnecessary neuronal connections, helping neurons refine their signals.

Oligodendrocytes These funny, fatty cells are like Beyonce's PR person; they help the message get out to the crowds. Oligodendrocytes send out processes that wrap around neuronal axons in layers, forming the brain's myelination. This fatty covering acts like insulation for the axon, helping neurons send signals faster throughout the body, making it possible for our brains to cram so many tiny cells in there while still sending signals fast enough to feel instantaneous.

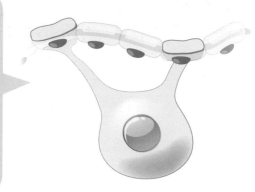

Epithelial Cells Meet the brain's bouncers! Because brains are so fragile, they need special protection against outside disease—so the blood vessels of the brain evolved to be extra specially difficult to get through, with what are called tight junctions keeping the cells very close together to keep outsiders from sneaking in and fewer leukocyte adhesion molecules that normally let immune cells from the blood into the body's tissue.

. . . and these aren't even all of the glial cells in your body! There are Schwann cells, which are like the oligodendrocytes but in your peripheral nervous system (outside of the brain and spinal cord), ependymal cells that line the ventricles of the brain and help produce cerebrospinal fluid, and satellite and enteric glial cells that support the sensory, parasympathetic, and sympathetic ganglia and the enteric nervous system.

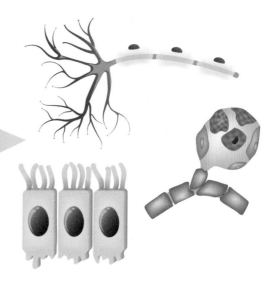

A WHO'S WHO OF EVEN MORE CHANGEMAKERS

Neuroscience and psychology history hasn't ended yet. Every day, new discoveries continue to transform how we think about and understand the brain, and fascinating people are bringing their A-game to the research table. Here are a few incredible scientists who are still changing the game for us all.

Huda Akil: In collaboration with her husband, Huda discovered the existence of endorphins—hormones produced by the brain that act like natural opioids. She also went on to show that in cases of severe stress, the brain can release endorphins to reduce pain. So all those action movies where the heroes just shrugged off those gunshot wounds weren't *entirely* inaccurate.

Carla Shatz: The first woman to receive a PhD in neurobiology from Harvard University, Shatz's research on the visual system of cats found that during development, neuronal connections aren't determined only by our genes—they also rely on neuronal activity to tell them which connections to maintain and which to throw out.

Elizabeth Loftus: A cognitive psychologist and expert on human memory, Loftus is largely responsible for demonstrating how easy it is to accidentally implant false memories, especially childhood memories, in individuals through suggestive techniques. This led to a complete overhaul of how many states handled the use of eyewitness and recovered memory testimony in the courtroom.

Doris Tsao: A pioneer in the field of systems neuroscience, Tsao's research using fMRI scanning on macaque monkeys found that individual cells in the brain respond to individual features of faces—a cell here for the eyes, a cell there for the mouth. Her team was able to pick this signaling apart so precisely that they could even reconstruct the face a monkey was looking at based just on the signaling of the neurons they were recording!

Daniel Colón-Ramos: Using *C. elegans* worms, Colón-Ramos studies the neurobiology of how neurons find and form connections (synapses) with one another, including their reliance on signals from glial cells. He's also part of a team building an atlas of the entire C. elegans system to provide a digital resource for other scientists to inform their research.

Kay Tye: Did you know Tye studies the neural circuits involved in emotion and motivation—using lasers? (No, seriously. See the section on optogenetics for more details). During her PhD, she found that the amygdala, which is important for processing emotions, increases its activity in rats when they're learning to pair a stimulus with a reward. Now she works on understanding how different neural circuits affect different kinds of behaviors and how they can be modulated to treat disease.

Damian Fair: In his study of autism, Fair uses fMRI to examine what he calls "connectotypes"—patterns of brain activity that are unique to individuals, like a brain fingerprint, which he says represent the distinct patterns of brain activity that determine how each person's mind works. Fair hopes to understand how these connectotypes are similar and different in autistic patients, which may clarify the underlying biology of the condition.

Bianca Jones-Marlin: With a keen interest in understanding how parenting affects development, during her PhD, Jones-Marlin found that the "love" hormone oxytocin could actually "turn up the volume" in the brains of female mice who were listening to baby mice cry: Non-mothers were more likely to respond to and care for pups when they were given extra oxytocin compared to those who weren't. So it's not that your mother didn't love you—you just weren't loud enough.

For a long time, we had no defined way of diagnosing mental health issues. The 1840 US census simply mashed all mentally ill folks into one category of "idiocy/insanity." Unfortunately, this statistic was useless, not only because the label was so poorly defined, but also because several census takers marked every African-American individual as "insane." Forty years later, the fledgling American Psychiatric Association developed the *Statistical Manual for the Use of Institutions for the Insane*. This guide standardized twenty-two different mental health diagnoses and, although its primary purpose was to get accurate statistics from mental hospitals, doctors started using it to diagnose patients. This became the predecessor of today's "mental health bible."

Mad World

In the wake of World War II, doctors started to notice that previously healthy, totally average men were now experiencing psychological issues after witnessing the horrors of war. Surprisingly, it appeared Freudian talk therapy techniques were a more effective treatment for traumatized soldiers than efforts to treat their condition as a physical health issue, which was the prevailing theory of the time. This totally shifted the view of psychopathology and helped distinguish mental health disorders from physical health issues.

By 1952, the American Psychiatric Association released the first *Diagnostic and Statistical Manual of Mental Disorders*. This first DSM covered 106 mental disorders and was actually intended to be used by clinicians for diagnostic purposes. If you are a healthcare professional in the United States, you have likely encountered the DSM (if you're not in the US, you likely use the ICD, which is similar and is put out by the World Health Organization). It is a compilation of descriptions, symptoms, and criteria used for diagnosing mental disorders as we understand them today. Since its release, the DSM has gone through multiple updates, often with significant changes in each new edition (the last version was released in 2013). The DSM, like other medical texts, is a living document. As we make new discoveries, conduct more research, and incorporate more diverse perspectives, the DSM continues to change. It is as imperfect as our evolving understanding of neurodiversity, but is still valuable and useful!

What The DSM Gets Right

The DSM is our best effort (in America) to properly classify complicated human behavior. It is a wealth of information that provides structure and a common language for clinicians and researchers to conceptualize client conditions. Without it, we'd all just be winging it! Recent iterations of the DSM have addressed critiques that it over-pathologizes people. Now it emphasizes the client's perspective by often requiring "clinically significant distress and impairment to functioning" to fulfill a diagnosis. And as a living document, the DSM is not the be-all and end-all of mental health. It continues to promote growth and points out where we have gaps in knowledge in order to encourage research in those areas.

What the DSM Gets Wrong

Most famously, the DSM used to classify homosexuality as a disorder. It was thought to be a "paraphilia" rather than a normal variance of human sexuality. Thanks to the relentless efforts of gay rights activists in the 1970s, this was removed. Up until the DSM-5, which was published in 2013, women had to miss at least three menstrual cycles before they met criteria for anorexia nervosa. This criteria excluded those who continued to have periods and caused the misconception that men could not have anorexia. Speaking of women, "hysteria" was a diagnosis in the DSM up until the 1980s. Hysteria was a catch-all diagnosis for women in distress due to the ancient belief that it had something to do with the uterus. The cure? An orgasm. And Dissociative Identity Disorder, formerly called multiple personality disorder, is still in the DSM, but it is by far the most controversial. There's a good chance the different personas are artificially created by the patient or are unwittingly caused by the therapist. It will likely disappear from the DSM soon! Although we've gotten mental health wrong at times, at least we change and grow as we learn more.

OVERMEDICATED

About one in six Americans takes psychiatric drugs. And don't get me wrong, they can be highly effective. But the rapid rise of certain psychotropic drugs has raised legitimate concerns about whether doctors are overmedicating their patients. In the 1950s, antipsychotics like Haldol and Thorazine seemed effective for treating schizophrenia. But emergency rooms and psychiatric hospitals came under scrutiny for overusing these meds on "unruly" patients so staff didn't have to deal with them. Today, antipsychotic overprescription is still a concern in nursing homes where they're used on "difficult" dementia patients, despite inconclusive evidence of effectiveness and increased risk of death. And even among children, there are questions around stimulants like Ritalin and Adderall being overprescribed just because little Johnny is a bit hyper. Over two million children receive ADHD medication, some of whom are preschoolers. Why does this happen? Well, psychiatrists, drug companies, and insurance companies all share the blame. Drug companies are permitted to advertise directly to consumers in the US. Reimbursements for medication are higher and easier to get than they are for therapy, which makes it an easier choice. And psychiatrists make significantly more money prescribing medication instead of doing psychotherapy (yes, they supposedly do therapy)! If we want to change this pattern, we need to change the financial incentive structure.

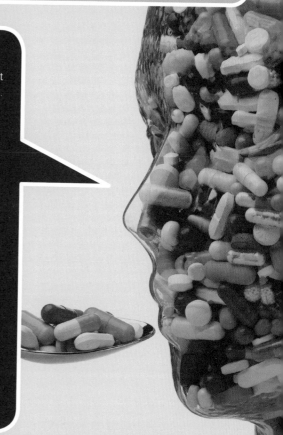

PART
TWO

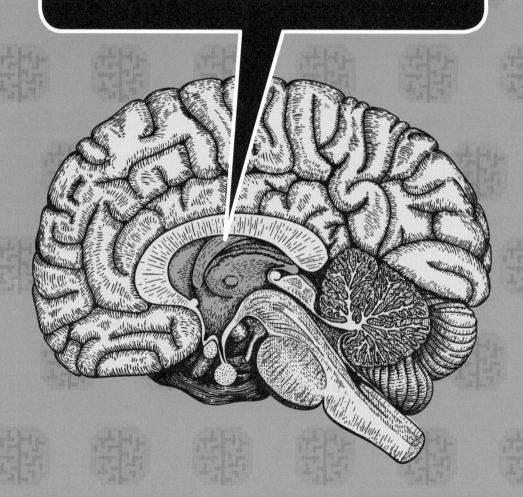

If you look at that big, beautiful organ between your ears, the midbrain is at the top of your brain stem, and it plays a part in overseeing lots of important bodily functions, like seeing, hearing, sleep, motor control, and temperature regulation.

In this book, we're using the term *midbrain* to refer to the middle section, where we're taking a step out of the past to focus on what we know about your brain right now, this minute, as you're reading this book, and what scientists know about how it all works.

In the coming chapters, we'll treat you to a "sensual" experience . . . by exploring the five senses—and maybe even one or two extras you didn't know you had. How does your brain translate light into colors and faces, sometimes even seeing faces and patterns that aren't there? What's the relationship between language thought? How does the human nose stack up against the rest of those of the animal kingdom?

We'll also dig into some deeper topics—sometimes literally, as we talk about structures located deep inside your brain—and start thinking about what it means to be human, with our big, squishy brains. What do scientists know about how and where memories are stored in the brain, and how do they know it? Why do we spend a third of our lives asleep, and why do we literally die without it? What is love (and why do we sometimes like it to hurt)?

None of these are easy questions to answer, but scientists are taking some pretty good cracks at it, and finding some interesting answers along the way. And try not to get too tripped up on how wild and weird it is that we're using our brains to figure out how our own brains work—let's just focus on what we've learned so far.

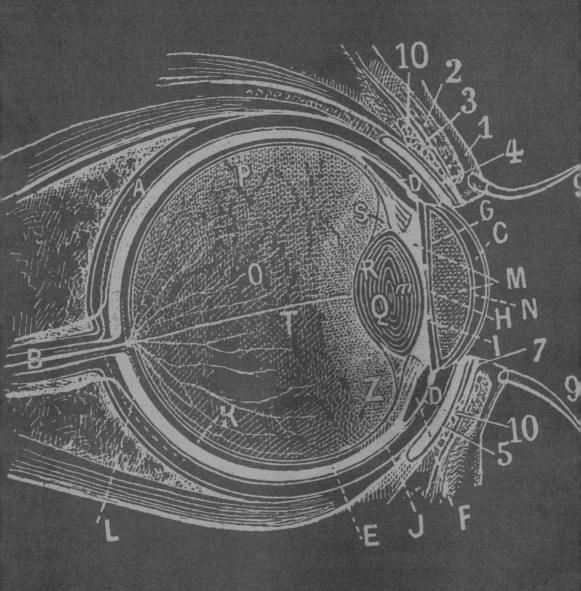

Chapter 8
I CAN SEE CLEARLY NOW!
(OR MAYBE NOT)

Vision is in the eye of the beholder; not every species has it, and even within a species, there's lots of variation. (Ever hear of colorblindness? Or prosopagnosia—better known as face blindness?) Even so, it's pretty much the most highly developed sense that humans have—or at least the one we understand the best.

Sight is such an incredible sense that Darwin himself felt that the evolution of the eye was almost impossible to explain by natural selection. But evolve vision did, and it brought us all the way from simple light-sensing organs to the complex and beautiful peepers that live in our own heads.

When light enters our eye, it passes through the lens and hits the back of the eye, where it triggers a cascading signal through the neurons of the retina to the optic nerve and then all the way to the back of the brain to be processed. Our brain breaks down the visual input into its component parts—colors, lines, and movement—and then rebuilds it into something meaningful using our memories and experiences.

Vision is one of the primary ways we interact with the world and gather information about it; with our eyes, we can drive a car, read a book, and play peekaboo with a baby. But it's not our only sense, and it's not a perfect system, which can lead to some pretty amazing adaptations!

NOW YOU GET THE PICTURE!

Vision starts in the eye of the beholder, and then travels all the way through your head for processing. Here's the pathway for how your brain turns those photons of light into mental photographs.

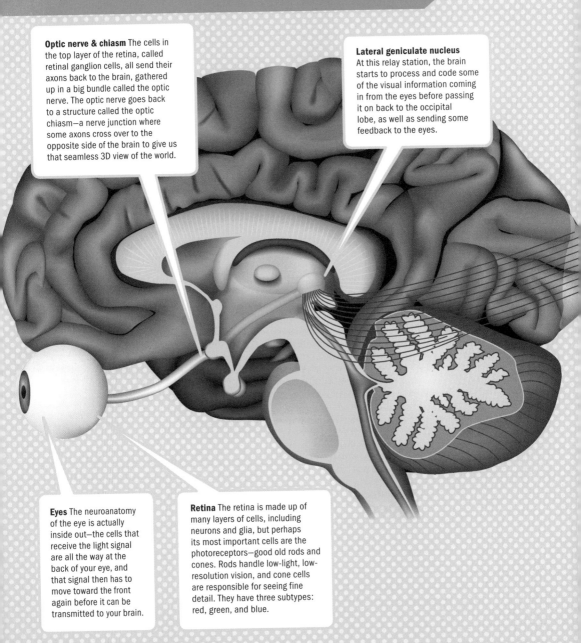

Optic nerve & chiasm The cells in the top layer of the retina, called retinal ganglion cells, all send their axons back to the brain, gathered up in a big bundle called the optic nerve. The optic nerve goes back to a structure called the optic chiasm—a nerve junction where some axons cross over to the opposite side of the brain to give us that seamless 3D view of the world.

Lateral geniculate nucleus At this relay station, the brain starts to process and code some of the visual information coming in from the eyes before passing it on back to the occipital lobe, as well as sending some feedback to the eyes.

Eyes The neuroanatomy of the eye is actually inside out—the cells that receive the light signal are all the way at the back of your eye, and that signal then has to move toward the front again before it can be transmitted to your brain.

Retina The retina is made up of many layers of cells, including neurons and glia, but perhaps its most important cells are the photoreceptors—good old rods and cones. Rods handle low-light, low-resolution vision, and cone cells are responsible for seeing fine detail. They have three subtypes: red, green, and blue.

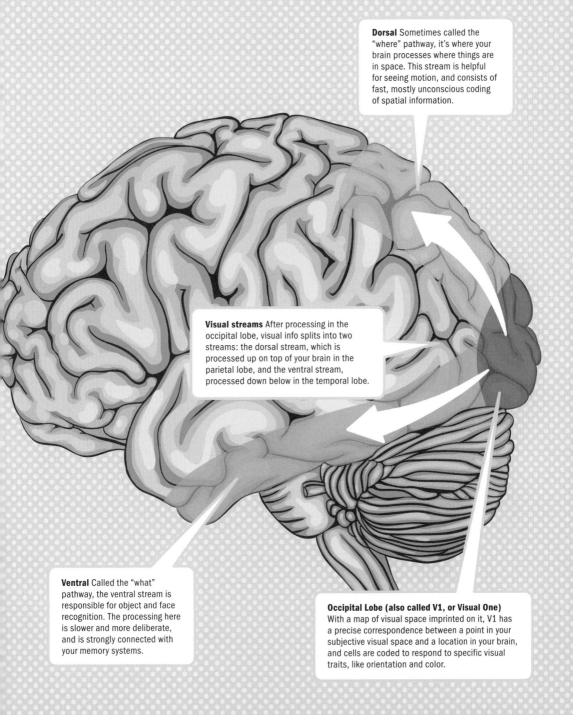

Dorsal Sometimes called the "where" pathway, it's where your brain processes where things are in space. This stream is helpful for seeing motion, and consists of fast, mostly unconscious coding of spatial information.

Visual streams After processing in the occipital lobe, visual info splits into two streams: the dorsal stream, which is processed up on top of your brain in the parietal lobe, and the ventral stream, processed down below in the temporal lobe.

Ventral Called the "what" pathway, the ventral stream is responsible for object and face recognition. The processing here is slower and more deliberate, and is strongly connected with your memory systems.

Occipital Lobe (also called V1, or Visual One) With a map of visual space imprinted on it, V1 has a precise correspondence between a point in your subjective visual space and a location in your brain, and cells are coded to respond to specific visual traits, like orientation and color.

COLOR VISION

She's Like a Rainbow . . .

Color vision is pretty incredible; thanks to just three types of cells in our eyes—those red, green, and blue cones—we're able to see all the colors on the visible spectrum. It might be more accurate to say that *because* we have only those three types, we're able to see *only* the "visible" spectrum, because some other species have all kinds of wild photoreceptors that let them see a kaleidoscope of colors beyond our own.

When it comes to photoreceptors, our neurons essentially use visible light as a neurotransmitter. Our photoreceptors contain light-sensitive proteins called opsins, and when those opsins absorb photons of light, it triggers a cascade of events within the cone cell that causes the neuron to fire a signal. Different kinds of cone cells have opsins that are responsive to different wavelengths of light; about 60 percent of the cells are red cones,

which respond to light at wavelengths around 560 nanometers (nm), and another 30 percent are green cones, responding to light around 530 nm in length. Only 10 percent of our cones are blue cones, with a peak response around 430 nm. All together, with our rod cells, the neurons of our retina can respond to light between about 400 nm and 700 nm, thus creating our visible spectrum.

. . . or Not

But what if you're not able to see all of those colors? What happens then? Certain genetic variations can mean that some folks are missing one or more of those types of cones, or their cone cells aren't working properly. This essentially leaves a gap in their vision, where some wavelengths of light aren't being correctly detected by the eye and so no information is getting sent to the brain.

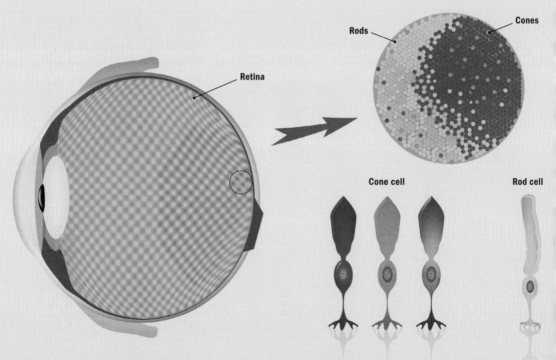

Rods

Cones

Retina

Cone cell

Rod cell

It's very rare for someone to be *completely* colorblind—that is, to have no functioning cones and thus not be able to see color at *all*. On the flip side, red-green colorblindness is the most common. This doesn't mean that people can't see the colors red or green at all—it's just that it's much harder for them to distinguish between the two colors. So, while most folks can pretty easily tell if an apple is red or green, folks who are colorblind might not be able to tell them apart without taking a bite first.

Trippy Visions

But now let's take it in the opposite direction—is it possible to see *more* colors? Well, sort of. Take the mantis shrimp and their sixteen photoreceptors (sounds like a glam rock band name, huh?) While having sixteen photoreceptors *sounds* like it should be a recipe for excellent color discrimination, that doesn't actually seem to be the case; instead, mantis shrimp can actually see *other* properties of light that humans can't see. For example, mantis shrimp can see UV light—so, I guess every day is like a rave party to them. But they can *also* see polarized light. You're probably familiar with polarized sunglasses, which work by blocking light waves vibrating horizontally, thus reducing glare off of surfaces like roads. Scientists think these shrimp use this ability to navigate within their environments, tracking the polarization of light to keep them going in the right direction.

COLORBLIND SUNGLASSES: HOW DO THEY WORK?

You've probably seen those feel-good videos of folks receiving some big, fancy, dark sunglasses and then immediately starting to cry when they put them on. These sunglasses are marketed as letting colorblind people more easily "see colors" by filtering out some wavelengths of light to create more separation between red and green hues. So do they work? Well, sort of. They essentially work by creating higher contrast between red and green hues so a colorblind person can tell them apart more easily—but they can't actually fully restore the dysfunctional cones causing the colorblindness in the first place, and because of the filters, the glasses actually remove even more color information from the light waves. And, not all colorblind people are colorblind in the same way, so what increases contrast for one person might actually make it harder to see color for another. What a colorblind person sees with these fancy, expensive sunglasses doesn't quite line up with what a noncolorblind person would see—but judging from those reaction videos, they do still have a pretty big effect on your vision!

FLYING BLIND

Outta Sight!

No, literally, let's talk about what happens when you lose your vision. We're generally pretty visual creatures, and rely a lot on our ability to see—but that doesn't mean we can't adapt to life without it. A person can end up blind from all sorts of causes; they may be born that way, due to congenital issues with their eyeballs or visual system, or may lose their sight due to disease or injury. But blindness isn't just a result of issues with your eyes; if you damage your occipital lobe, even if your eyes are still totally fine, you may lose some of your vision.

It's cool, though, because your brain is superpowered—seriously, you might end up like Daredevil. Because our brains have plasticity, even if you can't grow new neurons, it can still

adapt to losing sight and make some of your other senses even stronger as a result.

Say What?

People who are blind from a young age have been found to have excellent auditory discrimination—they can hear the differences between similar sounds much more easily than sighted people. That plasticity also means that your brain can sometimes repurpose the visual cortex to use it for other purposes—research has found that blind patients show activity in their occipital lobe in response to language and auditory information. Their sense of hearing literally creeps in to take over the visual cortex! This makes a lot of sense. Without vision, that occipital cortex represents a lot of prime real estate that your brain is going to want to use. And since our brains are good at reusing existing systems, why not move right on into that primo visual cortex?

Is There an Echo in Here?

We weren't kidding about that Daredevil thing. Some folks who are blind do literally use echolocation to help navigate their surroundings. Sound, like light, bounces on surfaces and can provide information about a space, like whether it's got a high ceiling or if you're near a corner. And some blind folks actually teach themselves to produce sounds, like clicks, to use that phenomenon to their advantage. They describe the resulting sensation as being similar to sight, providing them with actual spatial information about what's around them.

Studies looking at the brain activity of echolocators has found that this is accurate on a neurological level, too—when using their echolocation ability, the patients showed a lot of activity in visual brain regions as compared to auditory brain regions, indicating that they were actually using their visual machinery to navigate through physical space.

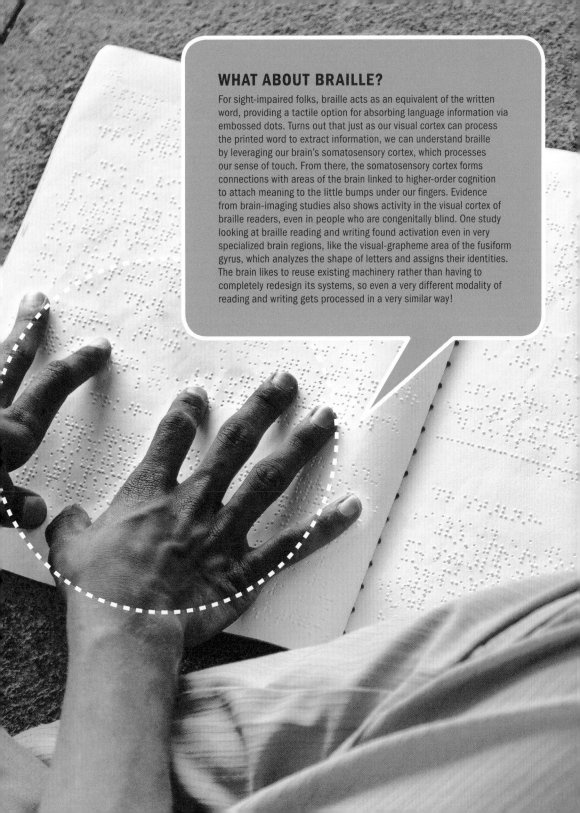

WHAT ABOUT BRAILLE?

For sight-impaired folks, braille acts as an equivalent of the written word, providing a tactile option for absorbing language information via embossed dots. Turns out that just as our visual cortex can process the printed word to extract information, we can understand braille by leveraging our brain's somatosensory cortex, which processes our sense of touch. From there, the somatosensory cortex forms connections with areas of the brain linked to higher-order cognition to attach meaning to the little bumps under our fingers. Evidence from brain-imaging studies also shows activity in the visual cortex of braille readers, even in people who are congenitally blind. One study looking at braille reading and writing found activation even in very specialized brain regions, like the visual-grapheme area of the fusiform gyrus, which analyzes the shape of letters and assigns their identities. The brain likes to reuse existing machinery rather than having to completely redesign its systems, so even a very different modality of reading and writing gets processed in a very similar way!

HALLUCINATION!

Are You Seeing What I'm Seeing?

Hallucinations can happen in pretty much any of our senses—sound, touch, taste, and so on—but perhaps the most often seen in media are visual hallucinations, usually as a result of consuming some psychoactive substance. But drugs aren't the only cause of visual hallucinations, and these odd phenomena have some pretty interesting underlying neurobiology.

Some visual illusions and perceptions get mistaken as hallucinations, like entoptic field phenomena. Basically, under the right lighting conditions, you can sometimes see objects *inside* the eye itself but if you don't realize what's going on, you might assume that you're hallucinating something weird. These include common things like floaters, caused by clumps of proteins or red blood cells moving around inside your eyeball, as well as blue field entoptic phenomena, which look like bright dots, caused by chunky white blood cells moving around in the blood vessels of your eye.

The Man in the Moon

Another visual phenomenon that's similar to a hallucination is pareidolia: our tendency to see shapes, patterns, or faces in inanimate objects. This includes things as innocuous as seeing a bunny-shaped cloud or the Man in the Moon, or conspiracy-theory fodder like the once-famous "Face on Mars." We're pretty sure the reason this happens is because our brains like to take shortcuts. It would be a lot of work for our brains to have to fully process and analyze all new visual information coming into the system, so the brain tends to fast-track information by applying things like patterns and easily recognizable shapes—especially faces—to whatever we're seeing. As a social species, our brains have *literally* evolved to quickly recognize faces and emotional states, so being able to see a face almost unconsciously is important—but it does mean that sometimes we get freaked out by rock formations on other planets.

What's Real? And What's Not?

A "true" visual hallucination is the result of a person perceiving some visual information in the absence of a stimulus; basically, the brain is making something up out of whole cloth. This can happen when you use a hallucinogenic substance such as LSD (more on that in a minute), but that's not the only cause; our brains are capable of hallucinating all on their own!

Some kinds of brain damage, like neurodegeneration caused by Lewy body dementia, and certain neurological conditions such as schizophrenia can lead to visual hallucinations, but so can ordinary circumstances like sleep deprivation or sensory deprivation. Even people who are sight-impaired sometimes experience visual hallucinations in a condition known as Charles Bonnet syndrome.

Visual hallucinations may be simple, manifesting as lights, colors, or geometric shapes, or more complex, with visuals including people, animals, and even complete scenes. From studying patients with dementia, it seems that there's a relationship between dopamine signaling in the brain and hallucinations; it may also be related to serotonin signaling in the brain,

since many drugs that induce hallucinations act on serotonin levels or receptors. You might think that hallucinations mean that the brain is overactive—as if the neurons of your visual cortex were sending too many signals and creating false visual information. But, studies using hallucinogenic drugs indicate that it could be that your visual system is *underactive*, and that the brain might be hallucinating because it's trying to fill in that missing information. Basically, your brain's just getting bored, and starts making stuff up to keep itself occupied!

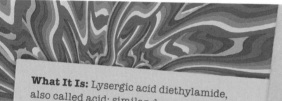

YOUR BRAIN ON . . .

LSD

What It Is: Lysergic acid diethylamide, also called acid; similar drugs include the 2C family, magic mushrooms, mescaline, ayahuasca, and others.

What Type of Drug It Is: Psychedelic.

What It Does: Also called hallucinogens, psychedelics are drugs that significantly alter your perception and may induce hallucinations and/or profound spiritual or religious experiences. Groovy, man. At low doses, LSD can cause mild euphoria and changes in visual perception. At higher doses, it can lead to full-blown visual hallucinations or deeply personal or spiritual realizations or experiences, and at extremely high doses, possible effects include time distortion, out-of-body experiences, and a total dissolution of one's sense of self.

How It Does This: LSD's active component structurally resembles our brain's naturally occurring transmitter serotonin. It's believed to bind to the brain's serotonin receptors and then sort of get stuck there, leading to the long trips most people experience (upward of 12 hours!) This induces a sort of overdrive serotonin response, and has cascading effects on other neurotransmitters, including dopamine.

What the Risks Are: LSD may play a role in acute psychosis in people with a family history of schizophrenia or other severe mental illnesses, but it's not considered particularly dangerous or addictive, and may actually be a useful tool for medicine. In early studies, a single dose is effective in reducing alcohol consumption in patients with alcoholism, and LSD has potential for treating anxiety, depression, and other addictions. Plus, you get to see pretty colors too!

Chapter 9

HEAR, HEAR!

Huh? What's that you say? Something about an earring? Oh, hearing! From our ears to our brains, the auditory system does a pretty amazing job of converting the physical vibrations of sound into information that we can understand, including spoken language—a major part of what makes us human!

How do our brains take the physical vibration of molecules in the air and turn it into an electrical signal in the brain? Well, it's surprisingly simple—kind of.

Sound begins as a wave, vibrating the molecules in the air around us, before it reaches our ears and is transformed into the neural signals that we process in our auditory cortex. Sound is all around us constantly—caused by everything from airplanes flying overhead to the clack of a keyboard to the purr of a friendly cat lounging in your lap. It gives us information about the world, and tells us about the things we can't see right in front of us.

But sound doesn't just come from our environment; it also comes from other people. And this is where sound has perhaps its biggest role in our lives: as language. In our tight-knit social communities, communication is key. Some scientists believe that language was critical for our success as a species by letting us connect more deeply with the people around us. That sounds like a big claim—but then again, here we are, using these words to connect with you!

AUDITORY PROCESSING

How does a sound get from a wave to a physical action in your body? Vibrations get transmitted and amplified through a complicated series of membranes, liquid, and bones in order to reach some neurons—it's a long way to the auditory cortex!

The Ear: Those funny folds of cartilage around the outside of your ear are called the pinna; they catch sound waves and direct them into your ear canal, which acts as a sort of amplifier for the waves.

Ear Bones: Malleus, incus, and stapes—oh, my! The three ear bones are all connected with one another, and act as levers to amplify the sound vibrations on their way through the ear.

Cochlea: Next, the vibrations get passed along to another membrane, called the oval window on the cochlea, whose name is Latin for "snail shell." This little snail-shaped structure is filled with liquid; when sound reaches the stapes, the bone vibrates and pushes up against the oval window, transferring that sound into ripples sloshing around in the fluid.

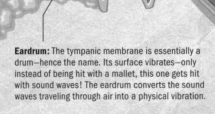

Eardrum: The tympanic membrane is essentially a drum—hence the name. Its surface vibrates—only instead of being hit with a mallet, this one gets hit with sound waves! The eardrum converts the sound waves traveling through air into a physical vibration.

EVERY TIME A BELL RINGS

Your parents weren't kidding when they told you that listening to loud music is bad for your ears. A big part of hearing loss from loud noises happens if you damage the stereocilia—if a noise is too loud, it jostles the hair cells too hard, and the stereocilia "flop over" and become nonfunctional. This can lead to hearing loss and tinnitus as you get older—but that's not the only cause of tinnitus. This ringing or buzzing in your ears can also be caused by damage, inflammation, or blockage anywhere between your ear canal and your brain. One of our editors had an artery compressing his auditory nerves, causing unbearable tinnitus that had to be corrected with brain surgery! You can create your own temporary tinnitus by tightening your jaw with your mouth open.

A HAIRY SORT OF CELL

Hair cells turn motion into what's called "action potential." How? Well, it's all thanks to some clever engineering by evolution. As the fluid inside the organ of Corti sloshes over the hair cells, it physically pushes on the spiky stereocilia on top of them—you know, the bits that look like hair. Imagine a gate with a spring on it—when all the hair cells are lined up, the gate is closed. But when you push on a hair cell, it pulls the stereocilia apart, which pulls on the spring attached to the gate, and causes the gate to fly open so the ions can get through!

Hair Cells: Inside the cochlea is the organ of Corti, which is lined with hair cells, so named because of their funny flat-top-haircut appearance. These cells get pushed by the fluid as it sloshes past, causing them to release neurotransmitters, which then signal the auditory neurons nearby to fire.

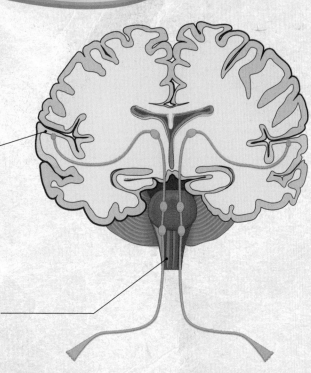

Auditory Cortex: Here's where all of that unpacked information starts getting processed for the first time. The auditory information is passed on to other places in the brain, including the frontal lobe, allowing you to put together all the various components of the sound and make sense of it.

Cochlear Nuclear Complex: The auditory nerve passes down to the cochlear nucleus in the brain, which then sends axons through several other structures where auditory sorting and unpacking occurs—each of these structures helps your brain break down the frequency, pitch, and localization of sound. Sound information then gets relayed through other brain structures to the auditory cortex.

KNOW THE LINGO

Language is a pretty key part of what it means to be human. We're really social, so we use language to communicate and connect with our fellow human beings. Amazingly, babies learn their primary language just by being exposed to it—it doesn't require any special teaching. And deaf kids who grow up without formal sign language will often invent their own "home signs" to communicate with family and friends.

Babbling Brains

Of course, language is an *extremely* complex behavioral system, so it's not very easy to figure out exactly where it's coming from. Back in the days when sophisticated brain surgery involved sticking an ice pick up someone's nose, understanding the neuroscience of language

was pretty much out of reach. As we learned with Broca and Patient Tan, some of our earliest understanding of language in the brain came from case studies of patients who had suffered brain injuries that damaged regions of the brain that were key for producing language.

Later on, the Wada test, which involved injecting patients with a barbiturate through the carotid artery to put half of the brain to sleep, showed us that language tends to be processed by a person's dominant hemisphere—usually the left hemisphere for right-handed people.

Split Streams

Humans mostly use speech to communicate with each other, so language is mostly an auditory ability. It turns out that a lot of the brain regions we use to hear sounds are also used to interpret speech. Some more recent neuroimaging studies led researchers to propose that language information gets split into what they called a dual stream model, where some language information goes up along the dorsal side of the brain on top, and some goes down along the ventral side.

The dorsal side of the brain includes regions like the motor cortex, so this dorsal stream is sort of the "speech" part of our system, serving to coordinate the complicated motor behaviors that let our mouths, lips, and tongues produce words. The ventral stream, on the other hand, passes information exclusively along the brain's dominant hemisphere, and it's what allows us to identify speech and understand what words mean.

Language on the Brain

Another reason studying language is so hard is because, well, how do you study a "purely human" trait in animal models? One solution: Find a species that has some parts of language, and study that species instead. Enter birds—despite having super-different brains, since our evolutionary branches split off before the dinosaurs walked the Earth, different bird species use unique songs to communicate and are studied as a model for complex vocabulary and grammatical processing.

WHY IS IT EASIER FOR KIDS TO BECOME BILINGUAL?

If you learned a second language as a young child, you had a serious advantage over adults trying it out the first time. This is due, in part, to brain plasticity. When you're born, you have an excess number of neuronal connections that make it easy to learn new information. But as you grow up, these connections die back and your neurons become less plastic, which means it's more difficult

to form new neural connections. By the age of 10, full fluency becomes significantly more difficult. Environmental factors may hold us back from bilingualism, too, including lifestyle changes and the fear of looking silly. But it's not too late! Even if you're an old fogey, there's no reason why you can't become extremely proficient. In fact, using two or more languages daily is good for your brain, and may even stave off age-related cognitive issues like Alzheimer's.

AM I EXPLAINING THIS RIGHT?

Language is such an important part of what it means to be human that many researchers think that the languages you speak can actually dramatically impact your perception of the world. This theory is known as the Sapir-Whorf hypothesis. For example, some languages, such as Mongolian, have distinct words for "light blue" and "dark blue," while other languages, such as Mandarin Chinese, just call it all "blue." While sighted people generally all have the same machinery for seeing the visible light spectrum, which means that our perception can't be entirely influenced by language, there is some evidence that having linguistic distinctions between shades can sharpen a person's perception of those colors. For example, in one study comparing Mongolian and Mandarin speakers, Mongolian participants were faster at finding a light-blue object on a dark-blue background than Mandarin speakers, indicating that they could distinguish between the two shades just a bit more easily.

MUSIC TO MY EARS

Your life has been shaped by music. Perhaps lullabies soothed you to sleep as a baby, or you feel your musical taste says something about your personality. Undoubtedly, you've listened to a song that gave you chills or maybe brought a tear to your eye. Music is a strange thing. It is not necessary for us to live, yet the love of music is universal. Every culture values and derives pleasure from it—but why?

My Pleasure!

This might sound cyclical, but we like music because it feels good. Human beings instinctively seek out pleasure; that's why we tend to indulge in sugar, sex, and drugs. When folks get wrapped up chasing these intensely enjoyable experiences to the detriment of their well-being, we call that addiction. Although there is no formal diagnosis of "music addiction," sex, drugs, and rock and roll all arouse the same parts of the brain.

Straight to the Dome

Music is extremely complex. In any given song, you have melody, harmony, rhythm, tempo, dynamics, timbre, and often lyrics. Songs can also conjure certain feelings or thoughts. To register all that, it's no surprise you end up using a lot of your brain, particularly the limbic system. You may recall from chapter 1 that this is a set of brain structures responsible for emotion, motivation, and memory. Within this system lives the brain's reward pathway, which, speaking scientifically, connects the ventral tegmental area to the nucleus accumbens to the orbitofrontal cortex. When we hear a song, the music activates this pathway, causing the ventral tegmental area to dump a bunch of dopamine. That dopamine gives us a euphoric feeling and may even cause goosebumps if the experience is particularly intense. Of course, music activates many more areas of the brain than just the limbic system, so it's no wonder music makes us feel alive!

You Say You Want an Evolution

While we understand the various ways that music affects the brain, we're still not totally clear on why music became pleasurable in the first place. Some say music may have been a method to attract a mate, while others say ancient peoples walked to a rhythm to hide their numbers, and still others theorize it was used to scare away predators. But the theory we enjoy most (which says nothing about its accuracy) is that it formed stronger social bonds by bringing communities together. Music has always been used in celebrations, to dance, to tell stories, and to share emotional experiences, for pretty much all of human history.

Music Soothes the Savage Beast

There is fervent debate over whether animals create music or not. At this point, it's too fuzzy to say either way. However, we have a somewhat better understanding of whether they appreciate human music. Some people like to turn on the radio when they leave their pets to go to work. After all, if humans like Beethoven, why not your cat? It turns out, though, that due to differences in their resting heart rate and vocal ranges, human music is not well tailored to other animals. But scientists have spent many hours and resources creating cat-specific music (as well as music for other animals like monkeys) and have found that . . . they like it!

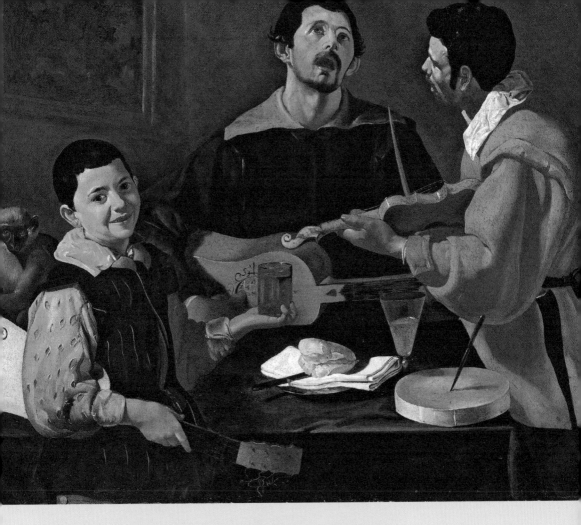

CAN MUSIC BE MEDICINE?

Sometimes therapy needs more than words. Sometimes you need to express it—in SONG!
Cue musical swell Music therapy first began during World War I for soldiers suffering
from "shell shock" and other mental health issues. When musicians were brought in
to play for injured veterans, doctors realized that soldiers who engaged with the music
recovered faster. It was healing them! You may have "prescribed" yourself some music in
order to relax or feel better—and it probably worked. Today, music therapy is an evidence-
based treatment conducted by trained professionals that involves playing or listening
to music as a way to access difficult emotions, engage in self-expression, or cope with
stressors. It is highly effective in treating symptoms of anxiety, depression, PTSD, ADHD,
schizophrenia, and autism—and, most significantly, music therapy seems to speed up the
recovery of individuals suffering from neurological disorders and traumatic brain injuries.

THAT SOUNDS OFF

Do You Hear What I Hear?

Not everyone *can* hear, for one reason or another. Deafness is an audiological condition that can be congenital (you're born with it) or acquired due to disease, injury, or aging that leads to a complete or nearly complete lack of hearing. Other folks may be hard of hearing to varying degrees.

Congenital deafness can be caused by environmental factors (for example, if a baby is exposed to certain kinds of infections or illnesses during pregnancy), but it can also be genetic. Sometimes these genetic variations affect other sensory systems or behaviors, such as Usher syndrome, which leads to deafness, blindness, and challenges with balance. Often, though, genetic deafness is nonsyndromic—the person shows no other symptoms.

Sometimes deafness or hearing impairment is caused by a problem with the physical machinery that translates vibrations into signals. For example, frequent ear infections can cause hearing loss because the buildup of fluid makes it hard for the membranes and ear bones to vibrate. Turning your headphones up too loud, or being exposed to loud noises in your environment, can also lead to hearing loss by damaging your hair cells so they fall over and can't move around in the cochlear fluid anymore and thus can't release neurotransmitters.

Different kinds of medication and other chemical exposure can also ead to difficulty hearing and, as a result, these substances are considered *ototoxic*. Obvious ototoxic compounds are things like lead or certain kinds of solvents, but there's also evidence that benign-seeming drugs such as ibuprofen and acetaminophen can cause hearing loss when used over long periods of time.

Losing the ability to hear can be difficult for some folks to deal with; noise-induced hearing loss can come along with other unpleasant symptoms such as tinnitus, and age-related hearing loss can increase feelings of loneliness among older adults when it makes it difficult for them to engage with others. But that doesn't mean deafness is a bad thing.

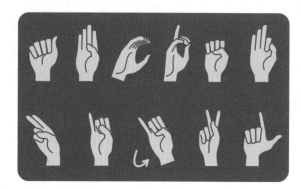

Deaf Punk

You may have seen *deaf* written with both a lowercase *d*, and with an uppercase *D*—What gives? In modern usage, lowercase-*d* deafness pertains to the audiological condition: a total or nearly total lack of hearing. Capital-*D* Deafness, however, is much more than that; it refers to a cultural community of people who use sign language as their primary language. Typically, these folks rely much more heavily on visual information and cues, including hand signs, for communication compared to people who use spoken language.

The Deaf community as a general rule works to promote a better understanding of deafness in the hearing world, and to emphasize that a lack of hearing does not need to be an impairment or a loss. They want hearing people to understand that with accessible resources, being hard of hearing or deaf is not a disability— it's just another way of living in the world that looks (and sounds!) a little bit different.

SCHIZOPHRENIA

There's a bad joke from the movie *What About Bob?* that goes, "Roses are red, violets are blue, I'm a schizophrenic, and so am I." A common misconception about schizophrenia is that it involves having multiple personalities. In reality, schizophrenia is a serious psychological disorder characterized by psychosis that often involves hearing voices, which makes it impossible to discern between what is real and what is not. Although the cause of schizophrenia is still poorly understood, the auditory hallucinations themselves seem correlated to reduced gray-matter volume in the superior temporal gyrus, which houses the auditory cortex. Hearing voices can be a super-scary experience: There could be one voice or multiple voices, one at a time or simultaneously, voices of acquaintances or strangers, and voices giving commands or shouting insults. But interestingly, the severity of these hallucinations can change depending on your environment. On average, people in the UK and the US perceive the voices as violent and hateful, and evidence that they're "sick." Meanwhile, folks in India and Africa are more likely to have positive experiences with the voices and do not seem as troubled by them. This perfectly demonstrates how mental health stigma and lack of acceptance from our communities can impact our clinical perspectives, causing more harm than good.

A BRAIN IMPLANT FOR HEARING!

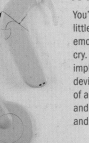

You've probably seen those YouTube videos of cute little babies "hearing for the first time," overlaid with emotional music and *definitely* designed to make you cry. They're showing the moment a person's cochlear implant is first turned on. These small electronic devices require surgery to be placed, and consist of an external microphone, a speech processor, and a transmitter to pick up and send the sound and a receiver under the skin. The receiver sends the signal to an electrode array within the person's cochlea, which then passes the message on to the auditory nerve and the brain. But these devices don't perfectly recreate natural sound, and are somewhat controversial within the Deaf community. "Curing" deafness with a cochlear implant is insulting for many who feel fulfilled and comfortable living without hearing, because it implies that deafness is a problem that needs to be fixed.

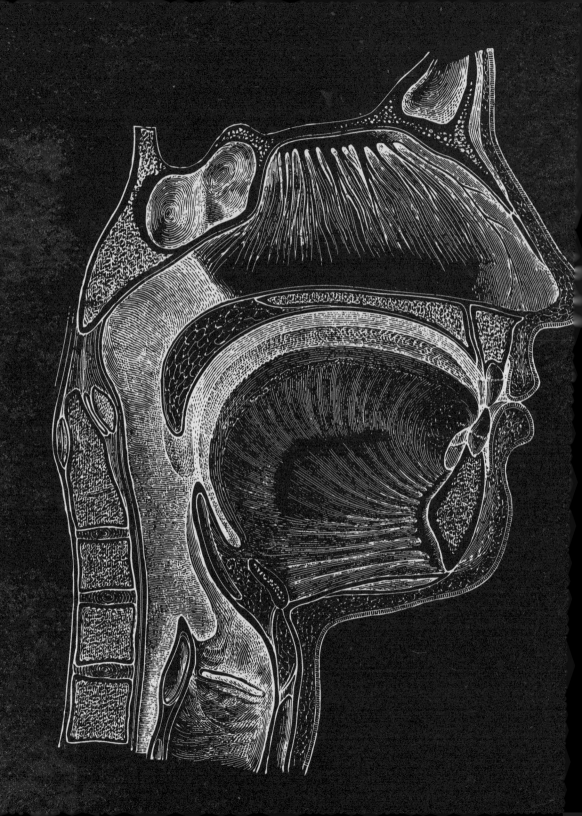

Chapter 10

YOUR BRAIN'S IMPECCABLE TASTE
(AND SMELL)

Why are we lumping taste and smell together in one chapter? Basically, because even though you have separate sensory systems to detect smells and tastes, both of them are detecting the same thing: chemicals. And when you eat, that smell and taste information all converges to give food its flavor. So, who are we to argue with your brain?

Arguably, taste and smell are the oldest sensory systems in the natural world. Both of them rely on chemosensation—the detection of chemicals in our environment. You taste things when they come into contact with your tongue, and you smell things when tiny molecules in the air called odorants bind to receptors in your nose. In both cases, a variety of unique molecules are detected by a multitude of receptors to trigger yet another cascading response through different brain regions so we can smell and taste the world around us.

The ability to detect molecules is nothing unique—even bacteria can do it, and likely have been doing so since life first arose on the planet. Detecting chemical signals in an environment would be critical for early life to survive, helping tiny single-celled organisms find food and avoid deadly habitats. Before we could evolve senses like sight, hearing, or even touch, we had to develop chemosensation.

But for humans, chemosensation has developed into much more than a survival instinct. Different tastes, and the foods that provide them, give us pleasure or cause us pain, and are inextricably linked to important cultural and historical events. Smell is deeply tied to memory, even when we can't describe a scent; I can't easily recall my great-grandmother's voice, but I do remember what her apartment smelled like. So . . . how does all of this work, and why are these senses so important?

A LICK AND A WHIFF

The sense of taste is also known as *gustation*. These days, we mostly tantalize this sense with fancy foods like chocolates and BBQ, or occasionally blast our taste buds with a Flamin' Hot Cheetos binge. But before humans had the opportunity to go to the grocery store and choose between three hundred varieties of sugary cereal, our ancestors used gustation to distinguish between tasty meals and deadly toxins. A refined palate would have been important for staying alive—now, it's important for staying popular at parties.

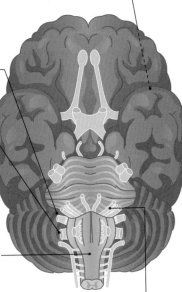

Olfactory Bulb: On the top of the inside of your nose, little olfactory receptor cells stick out into the nasal cavity, each expressing a single type of olfactory receptor. When an odorant binds, it starts a signal cascade that fires up to a bit of the brain called the olfactory bulb, sitting behind your eyes. There, many cells expressing the same olfactory receptor cluster together to connect with a mitral cell to pass the signal on to the primary olfactory cortex.

The Gustatory Cortex: After a run through the thalamus, as with most sensory information, finally, those good taste signals hit the gustatory cortex, located on an inner fold of the cortex between the insular lobe and the frontal cortex. Neurons in this brain region can respond to different tastes—sweet, salty, bitter, sour, and umami (savory)—as well as how strong the flavor is.

Glossopharyngeal Nerve: The last third of the tongue is innervated by the glossopharyngeal nerve, which also helps control swallowing.

Vagus Nerve: One of your ten cranial nerves, the vagus nerve goes down from the brain stem past the carotid artery to your guts, all the way down to the colon. It's responsible for most of the sensory information you get about your viscera —but it also has a branch, called the superior laryngeal nerve, that receives signals from taste buds in the back of your mouth and the esophagus.

Solitary Nucleus: All the nerves coming from the tongue connect to the solitary nucleus, a clump of neurons located in the medulla oblongata in the brainstem. This brain structure acts as a relay station for sensory information coming from the mouth, ears, and guts, as well as controlling critical reflexes like gagging and coughing.

Facial Nerve: About two-thirds of the tongue is innervated by the facial nerve, another cranial nerve. It also helps control facial expressions and saliva and tear production. Cue the waterworks!

Taste Buds: Each taste bud contains between 50 and 100 taste cells. Individual taste buds have tiny, wiggly little protrusions, called microvilli, that stick out into the taste pore, where they can come into contact with those tasty food bits.

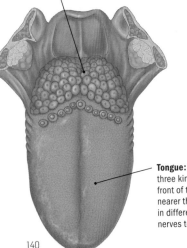

Tongue: On the tongue, taste buds may be found in three kinds of squishy structures called papillae; at the front of the tongue, you've got fungiform papillae, and nearer the back, foliate and vallate papillae. Taste buds in different areas on the tongue connect with different nerves to send signals to the brain.

TASTE BUDS

When I was a wee baby neuroscientist, I can remember learning about the newly discovered taste receptor, allowing us to detect umami, which is savoriness. Up until that point, we had been taught that we had only four tastes, not five —and that those tastes are carefully arranged around our tongues so we can taste sweet things at the front, bitter in the back, and sour and salty on the sides.

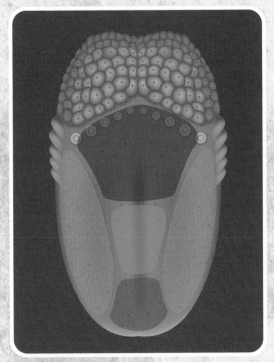

The Misconception

Kids are commonly taught about the "taste map" of the tongue, starting from a young age, but it turns out that taste buds don't work that way.

Why Did We Think That?

This was all based on a little misunderstanding of a German research paper published in 1901 (or so say those who can read German). The

researcher, David P. Hänig, was studying how much of a taste is required before different areas of the tongue can register the taste as a taste. Like, how much salt does the tip of the tongue need to recognize the taste of salt? His results indicated that different areas of the tongue has slightly different thresholds for registering tastes—but they did not show that specific areas of the tongue are responsible for detecting specific tastes!

The Real Story

Tasting things in different areas of your tongue isn't real; we do have only a limited set of taste receptors, but they're distributed all over the tongue, in all of our different taste buds. And

no matter where they're located, we know of five basic tastes, not four—salty, sweet, sour, bitter, and umami. Each flavor is the result of a different kind of chemical: Saltiness is caused by sodium or potassium ions, sourness is caused by acids, sweetness comes from sugar, umami is a result of glutamates, and bitterness . . . well, bitterness comes from lots of things. It's a critical flavor because, as proven by your angry ex still texting you five years later, bitter things are often pretty bad for you!

OLFACTION ACTION

You know, Bill Shakespeare was on to something when he said, "A rose by any other name would smell as sweet." The lovely smell of a rose has nothing to do with its name, but rather the chemicals it releases and the way your brain perceives them. And the smell of a rose isn't so simple, either: Your nose has hundreds of olfactory, or scent, receptors and can detect up to a *trillion* unique smells! But we're not so special. See how our nose stacks up to other noses!

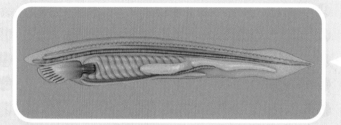

Some olfactory receptors go back millions of years—we share some scent receptors with **lancelets**, little fishlike critters whose evolutionary lineage diverged from ours over 700 million years ago. Lancelets don't have noses, though; instead, they smell with their skin.

The tiny nematode worm, *Caenorhabditis elegans*, is popular with smell scientists—because even though nematodes have only 302 neurons, 32 of them are dedicated to chemical detection. Seven percent of their entire genome is dedicated to genes that produce chemoreceptors!

The **turkey vulture**'s sense of smell is pretty keen—so keen, in fact, that these vultures are known to congregate around leaky natural gas pipelines. Not because they're big fans of gas—it's naturally scentless—but because of the additive ethyl mercaptan, a stinky compound that lets us know if there's a leak in the line and just so happens to smell like rotting meat.

Every year, **salmon** migrate from the ocean back to the streams where they were born to spawn and make new baby salmon. But how do they figure out which stream is home, sweet home? They follow their noses! As baby salmon, they imprint on the scent of their stream.

Have you ever noticed that sometimes **cats** stare at you with their mouths hanging open, looking goofy and shocked? Turns out they're not surprised by your appearance as much as they are by your smell. Often, cats making this particular face, called the flehmen response, are directing olfactory information to their vomeronasal organ—a separate scent-detector that humans have but don't use. It's often used for detecting pheromones, and has links to brain regions that direct aggressive and sexual behaviors.

We often think of **humans** as having a relatively weak sense of smell, and other animals like dogs as being "super sniffers"—but it turns out that humans are actually pretty good at smelling. This is another myth from everyone's favorite racist brain surgeon, Paul Broca, who claimed that the sense of smell is "animalistic" and thought humans could detect only 10,000 odors. In reality, with all the possible combinations of scent receptors, scientists think we can detect up to a trillion unique odors.

Who's the greatest smeller of them all? You were probably expecting to find dogs or maybe even pigs at the end of this list, but it turns out that the longest nose knows: African **elephants** are the animal kingdom's smelliest creatures. Or, at least, they have the most genes dedicated to producing olfactory receptors, comparatively speaking—a total of 2,000 such genes, twice as many as dogs, and five times as many as humans! Elephants probably use their prodigious sense of smell to do a lot of things, like finding food and identifying mating partners. But even though they might be able to smell more things than humans, that doesn't mean they can smell all of the same things as humans. Evolutionary analysis has found that while elephants have gained a lot of new olfactory receptor genes, they've also lost some of the genes that humans have. So, they can smell their mating partners—but they might not be able to smell a rose.

AH, YES, THAT FART TAKES ME BACK

Your sense of smell can act as a time machine. Catch a whiff of chocolate chip cookies in the oven, and suddenly, you're transported to your childhood home. Breathe in the scent of fresh-cut grass, and recall your summer job pushing a lawn mower. Get a sniff of chlorine in a hotel lobby and find yourself swept back to hot summer days spent by the pool. It's fascinating how smells can revive emotional, sometimes long-forgotten memories. How does that work? Well, when you smell a smell, what you're really detecting are tiny chemical particles. These odorants float into your nose and bind to olfactory receptor cells located way back behind your face. This then sends a signal to the olfactory bulb, where the signal gets processed and sent to two very important brain areas: the hippocampus and the amygdala. The hippocampus is primarily responsible for processing long-term memories, and the amygdala is in charge of processing emotional memories. None of our other senses have the same direct path to our brain's memory

centers, so it's no wonder smells can teleport us back in time to powerful moments in life. In fact, it's likely that you often experience the emotion of a smell before recalling the actual memory, because most of the information gets processed by the amygdala first before being sent on to the hippocampus. In a clinical setting, therapists take advantage of this memory pathway in treatment by using aromatherapy, which has been a practice for centuries, though you probably associate it these days with predatory essential oils multilevel-marketing pyramid schemes. And although essential oils will not cure cancer or autism, there is decent evidence that aromatherapy can reduce anxiety and depression symptoms, help individuals struggling with insomnia, and improve quality of life for people with chronic pain. It appears that the strong emotional recall of certain odors may have an effect on mood and alertness, giving clients a really easy tool for feeling more in control of their mental health. So, next time you're feeling stressed, follow your nose!

IN THE DUMPS AND CAN'T EVEN SMELL IT

The COVID-19 pandemic threw the world into disarray: People were getting sick, businesses were being shuttered, everyone was staying home, and we hardly understood anything about this virus. One of the most well-known symptoms of this coronavirus is the loss of taste and smell. Some experience a temporary dulling of these senses, while others fall victim to permanent olfactory dysfunction. Furthermore, research shows that a decrease in the sense of smell is the strongest predictor of depression in COVID-19 patients. Although sad, this isn't surprising. It's well known that impairments to olfaction can seriously affect one's quality of life—your food doesn't taste as delicious, you lose your appetite, you can't link smells to happy memories, and you may even experience relationship issues. No wonder sufferers feel depressed. But strangely, there is also evidence to support that depression may suppress your sense of smell. Yep, it turns out that there are common brain areas involved in both depression and olfaction, including the amygdala, hippocampus, insula, anterior cingulate cortex, and orbitofrontal cortex. So, which comes first? Well, sometimes it can be a real chicken-or-egg situation.

NOT SO SENSITIVE

You ever notice how some people just don't seem to notice certain smells? Like a person with a dozen cats might not realize that their house smells like . . . well, cat pee. Or a smoker doesn't know that their place reeks of smoke. This is because of olfactory adaptation, or olfactory fatigue—a totally normal phenomenon in which extended exposure to a smell wears out your nose and makes it much harder for you to detect it at all. It happens because you're actually tiring out your olfactory receptor neurons. Once you smell a smell, it kicks off a feedback loop that represses ion-channel activation and makes it harder for your olfactory receptors to respond at all. So, it's not that they're ignoring the smell—it's that they literally *can't* smell it.

TANTALIZING TASTE FACTS

It's Genetic, Not Generic

Even though we can all taste the same basic categories of flavor (bitter, sweet, salty, sour, umami), how we perceive those flavors isn't always the same, because we don't always have exactly the same taste receptors. Research indicates that different genetic variants in bitter receptors can affect how well we perceive bitter flavors. For example, depending on your genes, the compound PTC might taste like nothing, or it might taste super bitter. Your genes also explain why you might think cilantro tastes like soap—research has identified several specific genetic variants associated with tasting that soapy flavor.

She's a Super (Picky) Taster

Alie once dated a guy who ate his burgers well-done and totally plain—no ketchup or anything. She thought he was boring, but he may have just been a supertaster. Supertasters make up about 25 percent of the population and have more fungiform papillae on their tongue than the average person, and therefore more taste buds! Supertasters are especially sensitive to bitter flavors, but they also seem to respond more strongly to salty, sweet, and umami flavors. So, if you hate broccoli, it might be that you're not picky—you're just super.

It's a Miracle

So-called miracle fruit are actually berries from the *Synsepalum dulcificum* plant, and they contain a compound called miraculin. When you eat one of these berries, the miraculin in it binds to the sweet receptors on your tongue and makes them respond to sour flavors instead. This means that if you eat a miracle fruit and then suck on a lemon, it'll taste like a sweet, citrusy candy. Scientists are trying to leverage this compound to produce sweet-tasting foods that are lower in sugar.

THE SMELL OF SICKNESS

You've probably heard stories about how animals can "smell" sickness and death, like the dogs who can sniff out cancer or COVID-19. This is a real phenomenon, but how does it work? While our human sniffers aren't bad at smelling, we use them for very different things than most other animals. For example, we don't (usually) pee on things to leave messages for other members of the species. (We guess except for when people try to write their name in the snow.) So, a dog's sense of smell seems to be more sensitive to certain odors, which is why they're good at tracking people and detecting low blood sugar on someone's breath. We've also spent millennia training dogs to do our bidding, so we've gotten pretty good at asking dogs to discriminate between odors. All of this combined means that dogs are pretty sensitive to small changes in body composition and metabolism, and can be trained to respond to let their human handlers know that something is up.

HOW'S MY BREATH? NO, SERIOUSLY.

Everybody stinks, at least a little bit. And that's okay! Our bodies produce all kinds of secretions that give us some degree of body odor. Most folks aren't even all that aware of their own smell. But on the other end of the spectrum, some folks are really convinced that they're super smelly.

Similar in some ways to obsessive-compulsive disorder and body dysmorphia, there's a rare psychiatric condition known as olfactory reference syndrome where people are so convinced that they stink, they spend hours every day sniffing themselves, taking showers, and changing and washing their clothes. It's not known how many people have the condition because the hallucinations and obsessions can be so strong that people with olfactory reference syndrome often socially isolate themselves. The causes of the condition are unclear, but treatment with cognitive behavioral therapy and sometimes medications like antidepressants and antipsychotics can help.

EVERYONE (DIS)LIKES THAT!

Eating can be dangerous. We eat to survive, but many foods contain bacteria, parasites, or chemicals that could seriously hurt us. Luckily, rotting meat smells disgusting and poisonous plants taste bitter. But the opposite is true, too: We float toward the scent lines of fatty and sugary snacks like a hungry cartoon character. Evolutionary biologists think we evolved our sense of taste and smell to dislike things that may be harmful to consume (like poop) and to love high-calorie foods (like bacon). However, there's a lot of palate variability in humans. Every person's brain perceives tastes and smells differently. Your experiences, age, culture, socioeconomic status, hormones—heck, even your friends' opinions—can manipulate your sensory experience. So, just because you hate the smell of fish sauce or the taste of cilantro doesn't necessarily mean it's bad for you. But also, just because you enjoy the taste of Malört doesn't mean you'll make us drink it!

YOU SMELL ATTRACTIVE

The scent of someone you're attracted to can be . . . intoxicating. Those powerful smells and the, um, urges they can create might make you wonder if pheromones are responsible for turning you into a sex-crazed beast. But those animal instincts probably have more to do with regular ol' smells than with pheromones.

Animal Instincts

Pheromones are basically secreted hormones that can be detected by other members of a species, and they're not processed like most smells. In mammals, pheromones are detected by the vomeronasal organ and the information is transmitted to the accessory olfactory bulb, which is usually behind the regular olfactory bulb in the brain. In many species, male and female hormones provide important information about sexual status and mating receptiveness.

60 Percent of the Time, It Works Every Time

While humans do appear to have a vomeronasal organ, it seems to be pretty much unplugged. Our vomeronasal organs don't seem to connect to anything else, and while there's evidence for an accessory olfactory bulb in fetuses, by adulthood, it's pretty much gone. Plus, most of the genes associated with detecting pheromones in mammals aren't functional in humans. So, at the moment, there's not too much evidence for pheromones playing a role in human mating.

But that doesn't mean that humans definitely don't have pheromones! It's possible that we detect them with our ordinary olfactory system. And there's some evidence to suggest that differences in body odor and hormonal status can affect how attractive others find you—like a study where heterosexual women sniffed the dirty T-shirts of heterosexual men and preferred the sweaty pits of men whose genes were the most likely to produce babies with strong immune systems when combined with the women's own genes.

Chapter 11

SOME TOUCHING FACTS

Somatosensation is the sensory modality that lets us feel the world around us, telling us about the textures of surfaces, the motion of our body, the temperature of a surface, and more. So, put on your parachute pants, because all those other senses we already talked about . . . can't touch this.

Somatosensation may be our body's largest sensory system, with the greatest variety of sensations. It includes basically any sensation that tells your brain something about the way your body is interacting with its environment, *and* includes internal sensations. This might be why many folks argue that our sense of touch should actually be split up into a bunch of different senses, because somatosensation tells us information about so many distinct attributes: texture, vibration, weight, temperature, pain . . . the list goes on.

This big and complicated sensory system works as follows: sensors in your skin, muscles, joints, and viscera (that means your guts) are activated in response to some kind of stimuli, like running your hand over a soft cat's belly. Different kinds of information, coded by various kinds of receptors, get sent via these sensors (i.e., peripheral neurons) to connect with new neurons in the spinal cord.

For their part, the spinal cord neurons cross over to the opposite side of the brain and connect via the thalamus relay station before heading up to the primary somatosensory cortex—which literally contains a map of your body to detect and process all that incoming data. Read on to learn how our big, beautiful brains break down all of that information so we can understand it.

REACH OUT AND TOUCH ME

Somatosensation actually involves a lot of different kinds of sensations, responding to everything from vibration and texture to pressure and temperature. How does the body know what's what? It all comes down to the sensors.

Mechanoreceptors are perhaps the most classic kind of "touch" receptor; they respond to pressure and distortion on our skin, and different kinds of receptors in different layers of the epidermis have different responses.

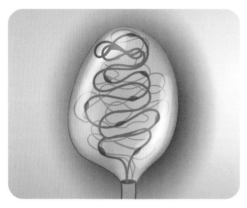

Meissner's corpuscles, or tactile corpuscles, respond to light touch and vibrations. The corpuscles themselves aren't actually part of neurons—instead, the neurons are wrapped around the corpuscles, ready for a signal. These are very sensitive receptors, for fine touch discrimination, like the brush of a feather.

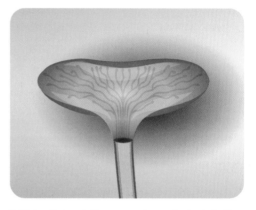

Ruffini endings, or bulbous corpuscles, are long, spindle-like nerve endings, located deeper in your skin and in the connective tissues of the body, and they respond to skin stretch and angle change in your joints.

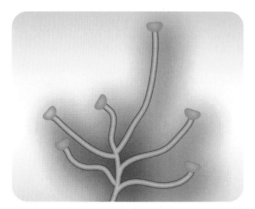

Merkel nerve endings, or Merkel discs, are somewhat like Meissner's corpuscles—the disks themselves are actually skin cells wrapped in axonal terminals. When subjected to pressure, the Merkel cells spit out serotonin, which activates the surrounding neurons. These cells help us detect ongoing pressure, like a hand holding yours.

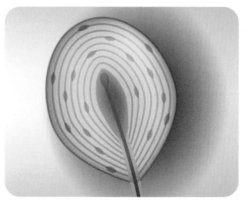

Lamellar corpuscles, or Pascinian corpuscles, are sensors shaped sort of like a layered onion found at the tip of a sensory axon. They're highly sensitive to vibration, like the purring of a cat, but filter out low-frequency stimuli, like sustained pressure.

Thermoreceptors detect hot and cold—and, interestingly, they're also the reason that mint tastes "cool" and jalapeños taste "hot!" The exact mechanism of temperature sensing is still a little bit of a mystery, but a family of sensors called transient receptor potential, or TRP, channels are a key part of it. These special ion channels can be activated by temperature changes, as well as by certain chemicals.

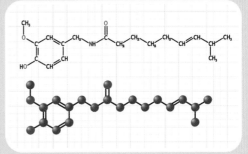

TRPM8 seems to be a primary sensor for cold temperatures in mammals, activating in response to temperatures below 20°C (about 68°F). They also respond to menthol, the chemical found in peppermint that also acts as a natural analgesic—explaining why mint produces a "cooling" sensation!

TRPV1 lets you know when things are getting uncomfortably warm, activating in response to temperatures over 43°C (109°F). They also respond to capsaicin—the chemical that makes jalapeños spicy—and to allyl isothiocyanate, which makes wasabi hot. These stimuli all lead to the painful, burning sensation we associate with high heat and very spicy foods.

Proprioceptors are what let us know where our body is in space to help us interact with the world around us. Proprioception is a sort of sixth sense, and it's why most folks don't usually have trouble touching their nose even when their eyes are closed. In addition to the inner ear and vestibular system, which provide information on our body's orientation and balance, we have specialized receptors to let us know what's up with our bodies.

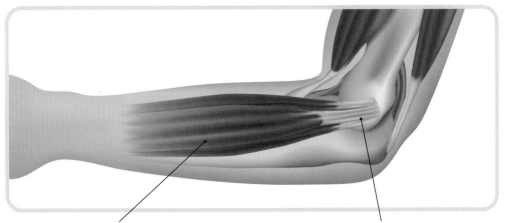

Muscle spindles are specialized nerve endings embedded throughout your muscles, and they respond to changes in muscle length and the speed at which muscle length changes, providing your brain with information about the position of your limb and its movement.

Golgi tendon organs are located in our tendons (duh). Tendons get stretched when muscles are contracting, and Golgi tendon organs respond when muscles are at their most active. These guys convey details about muscle tension, which translates to your sense of exertion.

CAN'T TOUCH THIS

Somatosensational

Once our bodies send all those signals up
to our brains to let us know what's going on
in the world, the information winds up in the
somatosensory cortex—a thick band on the top
of our brains where most of those mechanical
sensations like touch, pain, and vibration are
processed. In order to make sense of what
information is coming from what body part,
our brains evolved with an actual map of our
bodies spread out across the somatosensory
cortex. But of course, some of our body parts
are more sensitive to mechanosensation than
others—for example, our hands
are exquisitely sensitive and
capable of extremely fine
touch descrimination, while our
feet . . . not so much. So the
brain dedicates more of that
precious, wrinkly real estate
to the body parts that need to
have that fine-tuned sense of
touch, and less to others.

A Homuncu-what Now?

As a result, when scientists try to illustrate this
"map," things look a little . . . distorted. Enter
the sensory homunculus, a visual aid of a kind
that was developed by Wilder Penfield, Edwin
Boldrey, and Theodore Rasmussen. Based on
studies in real human patients who were awake
during surgery for epilepsy, the doctors were
able to stimulate parts of the sensory cortex and
figure out roughly what areas corresponded with
which body parts, and used that information to
mock up a homunculus (Latin for "little man").
The 2-D drawings look a bit odd, but not too
strange—they show that larger areas of
the cortex are dedicated to the fingers,
lips, and mouth, while smaller areas
are dedicated to the torso, legs,
and feet.

The 3-D homunculus,
however, gets a bit creepy
looking . . . like in this model
designed by artist Sharon Price-
James. While it does a very
nice job conveying information
about the relative innervation
of the body, we certainly
wouldn't want to encounter this
guy in a dark alley.

WHY ARE FEET SEXY?

Of all the various fetishes out there, perhaps the most common is the foot fetish. Many people, including Elvis, Ted Bundy, and Quentin Tarantino has found feet to be sexually arousing. This may seem strange—feet are one of our grossest, least-maintained body parts. There are questionable theories to explain this phenomenon. Freud thought people sexualize feet because they resemble penises (of course). Some think it's because feet are often the first part of the body with which toddlers play. Others say it's because bare feet are associated with sexually attractive occasions. However, the most likely theory for foot fetishism points to a neurological connection. Feet and genitals occupy adjacent areas of the brain's somatosensory cortex. Scientists believe that cross-wiring between those areas leads to foot-inspired arousal! In fact, some foot amputees who experience phantom limb syndrome report feeling sexual pleasure, or even orgasms, in their missing feet. Next time you give a foot massage, it could mean something more.

HEY, THAT TICKLES!

Have you ever stopped to think about how weird tickling is? Like, why on earth would our bodies evolve to make us laugh uncontrollably whenever someone lightly touches our neck?

There are actually two kinds of tickling—knismesis and gargalesis (Gesundheit). Knismesis is the very light sort of tickling, like what you feel if someone touches you with a feather. It doesn't really make you laugh, but instead feels like an itch. It sort of feels like a bug might be crawling on you, prompting you to brush it away—and in fact, scientists think this may be how this sensation evolved.

Gargalesis is the kind of tickling that makes you laugh, like when your friend won't stop digging their fingers into your armpits while shrieking "Gootchie gootchie goo!" Researchers looking at brain activity in volunteers who were being tickled this way saw enhanced activity in the hypothalamus, a brain region that plays a role in reflexive reactions like our fight-or-flight response, as well as the Rolandic operculum, which controls facial movements and emotional reactions. We likely can't tickle ourselves because our brain already knows what to expect, and so it can suppress the reflex.

Other animals can be tickled, too, including chimpanzees and even rats. Scientists think this may have evolved as a sort of submissive response, or maybe even as a way of helping juveniles learn how to fight off attackers without actually causing them pain.

A WORLD OF PAIN

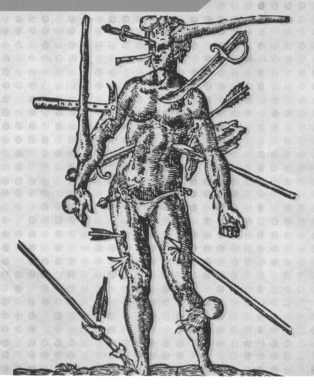

Pain is a real pain in the butt. And neck. And back. Of course, it's evolutionarily important—how else would you know not to touch that hot stove?—but sometimes the system goes haywire and you end up suffering for weeks from something as minor as sleeping a little funny. Trust us, we're over the age of 30. We know what we're talking about here.

Baby, Don't Hurt Me

Pain, like our other senses, relies on specialized receptors to let us know when a sensation is uncomfortable and possibly dangerous for us. Cells called nociceptors, found in your skin, your joints, and some of your internal organs, use temperature, pressure, and chemical stimuli to determine whether something is painful. The parts that actually sense painful stimuli are essentially nerve endings embedded in your

tissue, like the roots of a plant. Let's say you've just gotten a paper cut. There are three sorts of pain sensation associated with that event.

Ouch!

First, without thinking, you jerk your finger away in an immediate reaction. This is the result of a reflex arc, wherein a sensory neuron registers the painful stimuli and passes it along to your spinal cord. In the reflex response, the information isn't processed by your brain; instead, it's passed right back out of your spinal cord via motor neurons that control your movement, causing your hand to jerk back.

Pain in the Membrane

Next, you feel the first conscious "sharp" sensation of pain, as fast signaling nerves called Aδ ("A-delta") fibers respond, passing the signal to synapses with neurons in your spinal cord. This signal gets passed up to the somatosensory cortex, allowing you to consciously pinpoint the location of the pain. (Ah! My finger!) These fibers react only to mechanical and temperature stimuli, but not chemical ones, which might explain why you don't always feel the burn of lemon juice in a cut right away. Small blessings.

Slow and Steady Wins the . . . Pain Race?

Other nerve fibers, called C-fibers, react more slowly, passing along the deeper, "aching" pain. These cells project to the spinal cord as well, though they synapse with a slightly different set of spinal cord neurons, which project to the brain stem and the thalamus and then relay on up to the sensory cortex. These are the cells that'll respond to that nice lemon juice sting, and let it keep burning for a long time afterward.

Crossed Wires

Some of those nociceptors can be found inside our bodies, in our guts and other viscera—but our brains aren't always good at picking out exactly where pain information is coming from. We're not exactly sure why this is, but scientists think it likely has to do with the way that sensory information converges on neurons in our spinal cord and thalamus, and some information from our internal organs might get interpreted as painful sensations in other body parts. This is probably why one of the symptoms of a heart attack is pain in the left arm and, in women especially, in the jaw or back.

HIT ME BABY ONE MORE TIME

A little pain can sometimes be a good thing. And for some people, a lot of pain can be a *very* good thing. For generations, deriving pleasure from pain was considered a sign of mental illness, but these days, it's considered relatively commonplace—I mean, just look at how many folks got all hot and bothered over *50 Shades of Grey*.

The reality is that pain and pleasure often go hand in hand. Do you like to eat spicy food? Or how about that delightful soreness you get after a tough workout? Research in mice and humans has found evidence that our brains produce dopamine and endorphins in response to pleasurable *and* painful experiences, making both kinds of experience rewarding in their own ways. For some folks, this extends past jalapeños and into spanking.

From a psychological perspective, consent also plays a huge role; individuals enjoy giving and receiving pain in safe, defined environments, and many people report that BDSM experiences give them a chance to "give in" and fully experience their sexual desires, unrestricted by social norms and values.

THIS PAIN IS MENTAL

Sometimes pain doesn't come from a sinus infection or a hand on a burning stove. Sometimes pain comes from the mind. When a person experiences physical symptoms resulting from psychological causes or with no specific bodily cause, we call these *psychosomatic*. Everyone experiences psychosomatic symptoms to a certain extent—like when you're nervous and you get butterflies in your stomach or when you're embarrassed and your face flushes red. Essentially, experiences of emotional or mental distress are translated into very real physical effects—a process called somatization. But for some folks, the physical symptoms of psychological distress lead to persistent discomfort. When the pain doesn't go away, regardless of how it manifests, we call this a psychosomatic illness (or some call it a functional somatic syndrome). This umbrella term encompasses diagnoses like chronic fatigue syndrome, fibromyalgia, irritable bowel syndrome, and chronic tension headaches, to name a few.

What's the Meaning of This?

These syndromes are grouped together since it is believed that they have something in common due to their often-overlapping symptoms, including gastrointestinal issues, pain, fatigue, cognitive effects, and sleep difficulties. But our understanding of them is . . . fuzzy. Although functional somatic syndromes are fairly common, they remain something of a medical mystery. Since the symptoms have no apparent organic cause, psychosomatic illnesses are most often

diagnosed after ruling out all other explanations. We have not fully identified the mechanism of any of these ailments to date, but we have theories about what causes psychosomatic pain.

The Vicious Cycle

On the psychological side, research shows that individuals who have psychosomatic illnesses also report significantly higher rates of abuse, neglect, trauma, and childhood maltreatment. Many of these experiences threaten the individual's body, which is extremely stressful. Following this highly anxious event, it's believed, the individual becomes hypervigilant about how their body feels in order to avoid future bodily threats. So you can see how this works—if you're highly attuned to any somatic symptom, that symptom could elicit a stress response, which then produces more somatic symptoms, which then produces more stress, which then produces more symptoms—it's a never-ending feedback loop, and it becomes quite painful!

Axis of Evil

On the biological side, it is hypothesized that, following experiences of trauma or extreme stress, the hypothalamic-pituitary-adrenal axis falls out of whack. The HPA axis controls how you respond to stress, and regulates multiple body processes, like digestion, mood, and energy levels. It's believed that, in people who experience frequent or intense psychological distress, the HPA axis can become dysfunctional, creating irregular or lower levels of cortisol in the body. This change in cortisol results in the symptoms we see: gut problems, fatigue, anxiety, mood changes, and more. Worse yet, for folks with fibromyalgia and other chronic pain, researchers believe repeated nerve stimulation can cause the brain's pain receptors to develop a "memory" of pain, making the person more sensitive and reactionary to pain signals. No fun!

What It Is: Severe pain reliever.

What Type of Drug It Is: This category includes both natural and synthetic drugs, such as opium, morphine, heroin, fentanyl, Oxycontin, Vicodin, and codeine.

What It Does: Opioids relax the body and relieve moderate to severe pain. They can also treat coughing and diarrhea. However, due to the relaxed and euphoric "high" they give people, they are often abused or used recreationally.

How It Does This: Opioids activate opioid receptors (yes, we have receptors named for this stuff) located on cells in the brain, on the spinal cord, and all over the body. These receptors block the cells' ability to send pain signals to the brain and instead tell them to release large amounts of dopamine.

What the Risks Are: At high levels, opioids suppress brain activity that controls your breathing and monitors the amount of carbon dioxide in your blood. Opioids also have sedative effects, making you very drowsy and slowing down your breathing, resulting in a deadly combo that starves your brain and organs of oxygen. Every day, 128 people in the United States die from overdosing on opioids, and three times more Americans are addicted to prescribed opioids than to heroin. But just because you take opiates doesn't mean you will become a junkie. Like any other habit-forming drug, you need to moderate use, follow medical advice, and know your personal risk.

BUT IS THE PAIN REAL?

It's a common misconception that psychosomatic conditions are imaginary or "all in your head." Although psychosomatic symptoms may have mental or emotional roots, the pain is very real and requires treatment just like any other illness. It can be debilitating, negatively affecting the person's life to the point that their routine life is severely impacted. Unfortunately, a lot of patients are told by family or friends that they're just making it up. And some doctors may not be well versed in these disorders, may not know how to deal with them, or may not even believe patients, adding barriers to treatment. Luckily, the symptoms can be managed with counseling, physical therapy, diet changes, antidepressants, pain relievers, and other medications. Millions of people around the globe experience functional somatic syndromes, but they're not making excuses. It's just an invisible disorder that's poorly understood.

YOUR SIXTH SENSE  PROPRIOCEPTION

We've all got a sixth sense. And no, it's not seeing dead people. (Are you seeing dead people? You should get that checked out.) It's our sense of proprioception, or kinesthesia—our awareness of our body in space.

We already covered some of the receptors involved in this sense on page 151: muscle spindles and Golgi tendon organs. These specialized receptors are embedded in our muscles and tendons to respond to changes in muscle length and tension, letting us know where our limbs are in space, how fast they're moving, and how much force we're exerting with them. In addition to these special receptors, we have our amazing inner ear, and in particular our vestibular system, which plays a key role in our sense of balance and spatial orientation.

It's All in the Inner Ear

The semicircular ear canals, located next to the cochlea, contain their own set of hair cells—but instead of conveying information about sound information, these transmit information about movement, firing signals when our head moves up and down (like when nodding), back and forth (like when shaking your head), and from side to side (like when you touch your ear to your shoulder). This sends information to the brain about our rotational movement, helping us orient ourselves in space, especially when we're moving our head or our bodies.

Staying Balanced

Our inner ears also have special calcium carbonate structures called otoliths that help us sense acceleration and, especially, gravity. The brain combines input from these inner ear organs with visual data from the eyes to provide even more information—like whether we're turning our head or if the whole room is actually spinning—so we know what's going on with our bodies.

The information from the vestibular system goes to a bunch of places, including the spinal cord and thalamus, but one of the big targets is the cerebellum—that bit of the brain on the back that kind of looks like a plate of spaghetti. The cerebellum plays a huge role in movement control, and particularly in precision, balance, and coordination. Like most bits of the brain, the cerebellum is pretty important—but strangely, there have been cases of individuals who completely lack a cerebellum who get around just fine, if a little more wobbly than normal. That brain plasticity does a lot of heavy lifting!

HOW MANY SENSES DO WE HAVE, REALLY?

The standard five senses have been taught to kindergartners since time immemorial; you can thank Aristotle for that. But why do we only talk about five senses? After all, I definitely just told you all about at least one additional sense, proprioception. Really, how we split up our senses comes down to figuring out just how we categorize them. Our senses fall mostly across modalities—our visual system detects energy in the form of light (photons), our auditory and somatosensory systems detect mechanical stimuli (like vibrations in the air and touches on our skin), and our senses of taste and smell rely on chemosensation (detecting chemicals). If we like, we can split somatosensation up into multiple senses, like the senses of touch, temperature, and pain, or count internal sensations as their own kind of sense—like a sense of hunger. And some of the receptors we possess can detect more than one kind of stimuli—like the TRP receptors, which respond to temperature and to certain chemicals. So maybe we have dozens of senses. But most people would agree that we have at least six—vision, hearing, taste, smell, touch, and proprioception.

Touch is a powerful means of communication. It's the first "language" we learn as infants, and it's the first sense we develop in the womb. If a mother caresses their child as they fall asleep, we could describe how it activates the peripheral nerve fibers and yadda yadda yadda. But there is more information being sent through that caress.

First Contact

Indeed, there is a distinction between discriminative touch (sensing that you are being touched) and affective touch (touch that elicits an emotional reaction), and it turns out that humans are exceptionally good at identifying that extra information. In one study where participants were separated by a physical barrier, except for a hole large enough for an arm, researchers asked one person to communicate an emotion to the other person using only a one-second touch to their forearm. Emotions like compassion, gratitude, anger, love, and fear were correctly interpreted a majority of the time, showing how attuned we are to the emotion of touch.

No Touchy

Touch can also have a psychological effect. Humans are wired to be touched, but sometimes that need goes unmet. When we experience touch starvation (also known as skin hunger or touch deprivation), it can have serious, long-lasting effects, including feelings of loneliness, depression, anxiety, difficulty sleeping, and low relationship satisfaction. Affective touch can also suppress pain—a healing kiss on a scraped elbow makes it all better. But of course, not everyone enjoys being touched. Some people find warm hugs to be extremely uncomfortable and overstimulating. In neurodiverse individuals, like autistic folks, an ideal "touchscape" is vital to feeling stable. Some people may struggle with affective touch, resulting in classic symptoms. But if you have two consenting parties, tell someone you love them with a warm embrace. It's good for you.

Chapter 12
THAT REMINDS ME

Ah, yes, I remember the good old days. Or . . . were they good just because I remember them that way? Our memories are a huge part of what makes us who we are, so it might be a bit unnerving to learn that our memories aren't exactly the most reliable.

As we wander the world from birth to death, we hold onto past experiences, and use them to shape how we think and act. Memory lets us store all of that information, and then retrieve it again later on. Sometimes we take our memory for granted, but it's actually a complicated process that involves a tangle of different brain regions and some complicated biology. In neuroscience and psychology, we talk about memories being short term, meaning we retain the information for only a few minutes, or long term, where information is stored and retrieved over days, months, or even our lifetime.

Scientists split long-term memory into two big categories: declarative (or explicit) memory, and nondeclarative (or implicit) memory. Declarative memory is how we remember facts and events, while nondeclarative memory is responsible for habit formation and skill development, among other things.

This isn't just a way of categorizing different kinds of memory—they actually seem to be different, biologically. The hippocampus is necessary for retaining explicit memories, but not for implicit memories. And, memories appear to be stored in different places, depending on the kind of memory. For example, our conditioned emotional responses require the amygdala, our familiarity with our body comes from the cerebellum, and memories of facts and events rely on the medial temporal lobe, thalamus, and hypothalamus.

But don't go thinking that we really understand much about how memory works. Read on to build some new memories about how hazy and weird our memories really are.

THE FAULT IN OUR MEMORIES

Uh, I Forget

There are many explanations for why we forget things. Sometimes it's a problem with retrieving information—you're trying to remember that one actor in that one movie, but it's suddenly vanished from your mind. This may be due to decay theory: essentially, "Use it or lose it." If you don't retrieve and rehearse the memory over time, it fades and disappears. However, this theory doesn't explain how some older long-term memories (like all the times you've embarrassed yourself) remain incredibly stable, despite little recall. Some other scientists think your memories may compete and interfere with one another, meaning that either new memories are difficult to store, or they may even overwrite some older memories.

Of course, we don't store everything in long-term memory. Many details (like what you were supposed to pick up at the store) slip out of mind because your brain either deemed them unnecessary or because the encoding into long-term memory gets interrupted. And at times, our brains purposefully forget things, particularly traumatic or disturbing experiences, in order to reduce the pain of those memories. So, cut your brain some slack. It's trying its best.

Our memory sometimes seems totally accurate. It's almost as if we have a camera in our heads that records everything that happens around us and we're able to perfectly replay and recall the memories we save, even if they're a bit faded. But this is quite wrong. In fact, our brains are really, *really* bad at remembering things. Or, at least, at remembering one thing over another. You can lose years of calculus and trigonometry, but somehow remember the Preamble to the US Constitution. How does that happen?

IT'S DÉJÀ VU ALL OVER AGAIN

Now, let's examine the phenomenon of déjà vu: that strange feeling we sometimes get that we've lived through something before. Hey, wait a second—didn't I just say that? There are a lot of theories about why déjà vu happens, but it's likely not a memory from a past life or a glitch in the Matrix. But it *could* be a glitch in your brain! When researchers purposefully tried to induce déjà vu, they found that the medial temporal lobe lights up. This part of the brain is helpful in retrieving long-term memories, particularly of facts and events. Some scientists believe that a brief electrical malfunction in the temporal lobe activates your memory centers and, for a moment, makes you think, "Hmm, I have a memory of this." In fact, individuals with epilepsy who experience seizures in the temporal lobe often report having déjà vu right before a seizure.

I SWEAR IT WAS DIFFERENT

In 2010, a paranormal consultant named Fiona Broome thought that the South African anti-apartheid leader Nelson Mandela had died in prison in the 1980s. She remembered the news coverage and even the speech from Mandela's widow about his death. But the truth was that he was still alive (at the time) and had served as president of his country. This incorrect memory was an innocent mistake, no big deal . . . right? But strangely, Fiona noticed that many of her friends had the same recollection. And in fact, she found that thousands of others emphatically believed that they had witnessed Nelson Mandela's funeral, despite it never occurring.

It appeared that all these people shared the same fiction! How is that possible?! This phenomenon, known as the Mandela effect, demonstrates how susceptible we are to collective false memories. It's believed that these kinds of mutual myths form either because we're exposed to media that gets misinterpreted and confabulated, or because our brain is trying to "fill in the gaps" with information that seems to be the next logical step. In a broader world, it really shows how powerful misinformation can be! To show what we mean, here are a just a few of our favorite examples of the Mandela effect.

A Stain on Their Reputation
Hate to break it to you, but the loveable Berenstein Bears are actually the Beren*stain* Bears.

Curious George's Tail
Would you be surprised to learn . . . he never had one? Straaaaaaange.

Movie Magic
A lot of people remember the comedian Sinbad starring as a genie in a movie called *Shazaam*. But no, there is no such movie.

The Disappearing Basket
The Fruit of the Loom logo is only fruit. There was never a cornucopia behind it!

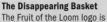

That's Not True . . . That's Impossible!
At the climax of *The Empire Strikes Back*, Darth Vader says, "Luke, I am your father." Except he never says "Luke" at all.

BEARING FALSE WITNESS

I Saw It All, Officer!

When you witness a crime or accident, authorities may come to you for details of what happened—and that testimony can often be a determining factor in insurance claims, lawsuits, or even sentencing decisions. The criminal justice system depends on evidence, including witness testimony, to convict those accused of a crime. But it turns out that there are so many factors influencing a person's memory that eyewitness testimonies are reliably unreliable.

Eh, I've Got the Gist

Serious incidents are typically stressful. Stress can be extremely helpful in keeping us alert and attentive. But when that stress gets too high, it can actually make you worse at remembering details. As mentioned earlier, our memory doesn't work like a video camera. When we recall memories, we don't play back a recording of what we saw. Your brain would explode if it tried to meticulously record the constant barrage of information every moment of life throws at you. So instead, your brain kinda ignores most things and just tries to remember the relevant stuff.

There Are Three Sides to Every Story

This is where problems happen. When you are later asked to recount the robbery, you'll remember some of the big details, and then your brain will do its best to fill in the rest of the picture with information that makes the most sense to you. And every time you're asked to recall that memory, it'll change slightly. It's almost like playing a game of telephone with your brain! Sometimes bystanders can't even remember basic facts about what happened. In crimes where someone is armed, eyewitnesses will often become so focused on the weapon that they fail to remember other important details—like what the person holding the weapon looked like. So, if you ever find yourself on a jury where the evidence is a single eyewitness's testimony, demand better evidence.

LIAR, LIAR, PANTS ON FIRE!

Some folks blur the line between truth and falsehood so much that it becomes almost impossible to tell what's real and what's not. These pathological liars tend to tell stories that depict themselves favorably, appearing either as the hero or the victim, but have no clear motive for lying. Pathological liars may even convince themselves of their own deceptions. Sound like any politicians you know? It's normal to defensively lie to avoid harm, but it's bizarre to fabricate stories for no personal gain. The exact cause is unknown, but it's associated with a chaotic home environment growing up. Pathological lying is not a stand-alone diagnosis, but rather is a symptom of personality disorders like antisocial, histrionic, and narcissistic personality disorders.

PICTURE-PERFECT MEMORY

Photographic memory seems like a superpower. Just imagine—you'd ace every test, remember every phone number, and never, ever get lost. While the term "eidetic memory" is sometimes used interchangeably with a "photographic memory," they're not quite the same: Eidetic memory is the ability to recall clear snapshots of specific memories, while photographic memory usually refers to remembering pages of text or strings of numbers. But is this superpower real life, or is it just fantasy?

Real Life

It does seem like eidetic memory is a real phenomenon, at least to some degree, though it's really only found in very young children. Kids with eidetic memory—"eidetikers," as they're sometimes called—can recall a visual memory in such detail and clarity, it's almost as if they are still looking right at it. In many cases, these memories are not just visual, but also include other sensory modalities, like remembering how the scene sounded and felt.

But, like we just said—this phenomenon occurs only rarely in very young kids, and even then, the memories aren't perfect. They're subject to the same kinds of distortion and misremembering as any other kind of memory—they just seem to be more vivid.

Will You Remember Me?

That more famous ability to remember pages of text or strings of numbers instantaneously? Yeah, probably not real. A lot of the hype around this idea came in the 1970s from one researcher named Charles Stromeyer, and his then student and later wife Elizabeth, who could reportedly memorize and then later combine random dot patterns to form mental 3D images (like a Magic Eye, but just in your brain). Elizabeth has famously refused to repeat the experiment with other researchers, and searches for others with similar skills have identified only one or two savants who are very good at remembering very specific kinds of information.

Enter My Memory Palace

This doesn't mean that memory superpowers don't exist at all. With the use of mnemonic devices and cognitive strategies to aid recall, the human memory is pretty ridiculously powerful. These tools add in extra layers of coding to any information you're trying to remember, to help your brain keep all the dots connected and make the information easier to recall. This is why you can probably still hear "My Very Educated Mother Just Served Us Nine Pizzas" and remember exactly what it stands for. Or what it used to stand for. RIP, Pluto.

WHERE DO MEMORIES LIVE?

While we know that certain brain regions are important for retaining different kinds of memories, it's less clear exactly how those memories are being stored. The theory with the most support so far is the synaptic theory, which posits that when we learn something, we're reactivating the same set of synapses over and over, which strengthens the connections between the neurons. This would mean that memories are essentially stored as networks of connected neurons in the brain. And the evidence is mounting! In a 2016 study at MIT, scientists were able to "tag" the neurons in the hippocampus that were activated while mice were learning about a new environment. Later, when the scientists activated just those same neurons, they were able to trigger the same behavior—even though the mice weren't in the same environment.

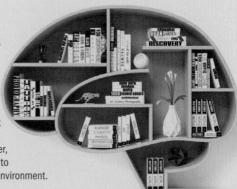

SWEET DREAMS ARE MADE OF THIS

Anyone who's ever had a teacher tell them to make sure they get a good night's rest before a big test is probably familiar with the idea that sleep is important for helping us consolidate our memories. But biologists didn't realize that sleep is critical for helping store memories until about a hundred years ago—and how it contributes is still kind of unclear.

Good Night!
Right off the bat, we know for sure that sleep deprivation makes it hard to pay attention and really focus on a lesson or a task, which can keep you from absorbing information in the first place. Even one bad night can have a measurable effect on your working memory—a short-term kind of memory that helps guide attention and decision-making.

Sleep Tight!
We know that sleep is important for memory consolidation thanks to research that started in the 1920s. A group of researchers taught subjects a list of nonsense words that they would never have encountered in the real world, to be sure that the participants were all learning new information. People who got a good night's rest were better at remembering the list than people who didn't get to sleep between learning the list and taking the test.

This research sparked a wave of studies that found a strong connection between sleep and certain kinds of memory—particularly, the retention of declarative memory. (That's facts and events—but you remembered that, right?) So, the teachers who tell you to sleep before your test are right—it's really important for remembering facts and figures.

Let No Bad Dreams Come to Fright!
Scientists have also discovered that in order to consolidate those declarative memories, you need a particular type of sleep called non-REM sleep—that is, non-rapid eye movement sleep. Even though you might relive your classes in your dreams (naked, even), it's actually when you're not dreaming that your brain is hard at work solidifying those memories. During this so-called slow-wave sleep, oscillating patterns of brain activity are thought to help strengthen the connections between neurons to make those memories really stick.

MY OWN LITTLE WORLD

You know by now that our memories are not perfect; we're prone to minor misremembering of all kinds of things, and it's almost absurdly easy to implant false memories. So, what's the deal? Why doesn't our brain encode all that stuff like the powerful computer it's supposed to be?

Memory Dump

Well, the world is a big, complicated, noisy place—lots of stuff is happening around us all the time. Our brains just aren't big enough or fast enough to process it all at once. So, our brains kind of have to be selective. We have to focus on just some of the information in order to really absorb it. And not all the information our brain processes gets encoded as long-term memory; some of it gets dumped out as soon as you close that book or walk away from that conversation. So, whenever you're recalling a memory, you're probably not recalling it in crystal-clear detail; you're just remembering the key information that you need to accomplish whatever you're doing.

Living in the Matrix

What if I told you everything you know to be true is only . . . maybe kinda true. Your brain takes the information it gets from its environment, filters it, and compares it to everything else you've ever experienced in order to help you understand exactly what's going on. Every single new experience gets scrutinized through the lens of your personal memories and your current consciousness. And, that consciousness is inherently biased. In fact, our brains will work to justify our feelings and actions to protect our own self-concept, which can in turn influence how we remember interactions and events. Like, if I think of myself as an animal lover, I might not remember that I killed a mouse that one time.

In essence, there's no such thing as an objective experience, because every sensation or thought or emotion you have is being filtered through your own unique reality. Heck, we even experience input delay! Even though our neurons fire really quickly, they're not instantaneous—so, everything you see and hear has already happened. Each one of us is living about 80 milliseconds in the past! Hearing all this, you could almost argue that each of us lives in our own personal virtual reality, built by our brain. (Keanu voice) *Whoa.*

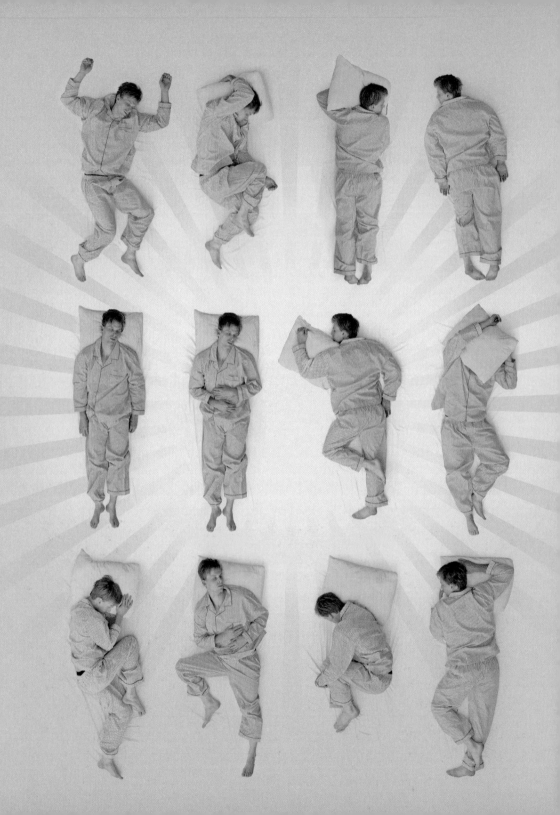

Zzz

Chapter 13

SWEET DREAMS!

Sleep eats up around eight hours of every day of your life. That means that you spend *one-third* of your life horizontal, trying to catch some shut-eye, utterly oblivious to the world around you. So, what gives? Why did evolution give us this weird requirement?

While our six-year-old selves might beg to differ, sleep is amazing—albeit pretty strange. All humans do it, as do most other animals. Sleep exists as a distinct state that differs from waking, one where we're mostly unresponsive to external stimuli, and includes the peculiar phenomenon of dreaming.

When asleep, the brain alternates between two unique substates: rapid eye movement (REM) and non-REM sleep. These substates cycle every 90 minutes, repeatedly through the night, and each substate is characterized by different types of brain activity. During non-REM sleep, the brain ripples with waves of neuronal

activity, and while movement is possible, dreaming is rare. REM sleep, on the other hand, is where the action is at—while you spend less time in this stage, it's also the sleep stage in which you dream.

But sleep is good for more than just dreaming. It's critical for properly encoding long-term memories, it bolsters our immune system, and (we think) it lets our brain go into "dishwasher mode" to clear out gunk between our brain cells and make sure everything is squeaky clean for the next day. So, seriously, catch some Zs. It's pretty important! Now, if you'll excuse us, we're going to take a nap.

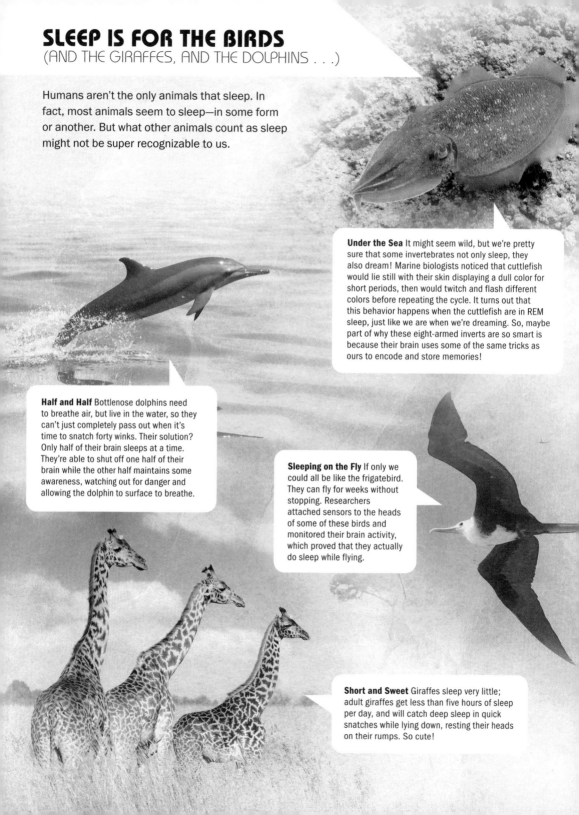

SLEEP IS FOR THE BIRDS
(AND THE GIRAFFES, AND THE DOLPHINS . . .)

Humans aren't the only animals that sleep. In fact, most animals seem to sleep—in some form or another. But what other animals count as sleep might not be super recognizable to us.

Under the Sea It might seem wild, but we're pretty sure that some invertebrates not only sleep, they also dream! Marine biologists noticed that cuttlefish would lie still with their skin displaying a dull color for short periods, then would twitch and flash different colors before repeating the cycle. It turns out that this behavior happens when the cuttlefish are in REM sleep, just like we are when we're dreaming. So, maybe part of why these eight-armed inverts are so smart is because their brain uses some of the same tricks as ours to encode and store memories!

Half and Half Bottlenose dolphins need to breathe air, but live in the water, so they can't just completely pass out when it's time to snatch forty winks. Their solution? Only half of their brain sleeps at a time. They're able to shut off one half of their brain while the other half maintains some awareness, watching out for danger and allowing the dolphin to surface to breathe.

Sleeping on the Fly If only we could all be like the frigatebird. They can fly for weeks without stopping. Researchers attached sensors to the heads of some of these birds and monitored their brain activity, which proved that they actually do sleep while flying.

Short and Sweet Giraffes sleep very little; adult giraffes get less than five hours of sleep per day, and will catch deep sleep in quick snatches while lying down, resting their heads on their rumps. So cute!

DO SHEEP DREAM OF BLEATING HUMANS?

If you have a pet cat or dog, you've probably seen them dreaming—twitching and maybe whining in their sleep, or even (as one particularly hilarious YouTube video demonstrates) sleep-bumping into walls. Do animals really dream? So far as we can tell, the answer is yes! We know that many species experience REM sleep, the stage in which humans dream. And research on brain activity in sleeping rats indicates that, at least some of the time, they're reactivating the same neurons they used to complete a maze task. Researchers think this means they're solidifying those memories—so, they may be running through mazes in their dreams!

FEEL THE (CIRCADIAN) RHYTHM

Whether you identify as an early bird or a night owl, you probably still experience regular cycles of sleep and wakefulness, and usually that means sleeping when it's dark and being awake when it's light. These cycles tend to follow regular patterns that align with the 24-hour rotation of the Earth, and are directed by internal processes collectively known as the circadian rhythm.

Rhythm of the Night

The theory about why these rhythms exist is that they essentially let us predict and take advantage of the natural cycles of our environment—the body is trying to optimize when it does certain things during the course of a 24-hour day. Many bodily functions are affected by the circadian rhythm, including sleep/wake cycles, digestion, hormone release, and body temperature. But we're not the only ones that follow these kinds of patterns. Circadian rhythms are found in almost all animals—and even found in bacteria!

A Well-Oiled Machine

In humans, this biological clock is controlled by a pacemaker located in the suprachiasmatic nucleus (SCN), a tiny brain region found within the hypothalamus. In many ways, the SCN acts like an orchestra's conductor; it takes cues from the environment—in particular, daylight and darkness thanks to direct input from the

eyes—and tells our body what "time" it is: time to eat, time to sleep, and so on. It does this by cycling the production and degradation of different proteins over a 24-hour period, creating feedback loops to regulate bodily functions during the daily cycle.

Blue-Light Blues

You know you shouldn't be looking at Twitter right before bed—and not just because doomscrolling is bad for your mental health. Since your brain uses cues from the light in the environment, staring at screens late into the night can disrupt your normal circadian rhythm, which is bad for your sleep. Ordinarily, as it starts to get dark out, the brain produces melatonin, a hormone that makes us feel sleepy (it's even a common over-the-counter sleep aid!) But evening exposure to blue light, whichin nature is produced mostly by the sun, suppresses melatonin production, so we don't get sleepy when we should, and we don't fall asleep as easily as we ought to.

Throwing that all-important circadian rhythm of ours out of whack can affect a lot of other bodily functions, like hunger and digestion, which can lead to other health problems down the road. So, if you're up late at night to, say . . . oh, I don't know . . . finish a book about the brain, consider using apps that bump up the warmth of your screen or wear blue-light-blocking glasses in the evenings.

JET PAST THAT JET LAG

When your circadian rhythm gets severely interrupted—like, when you fly across several time zones—you can end up with jet lag, leaving you lying painfully awake at 3 a.m. in your European hostel and passing out before the locals have even started eating dinner. Your body can't adjust to the new schedule instantaneously ,because those circadian rhythms are pretty entrenched, even if the sunlight outside has changed. So, your body will still send wake-up and go-to-sleep signals at the times it's used to, instead of the times it should. You can help avoid this by starting to adjust your sleep schedule ahead of time and forcing yourself to stay up until at least 10 p.m. when you arrive in your new location. That way, you can spend your evenings out on the town enjoying your vacation instead of snoring away in bed.

YOUR BRAIN ON . . .
METHAMPHETAMINE

What It Is: Methamphetamine, aka meth, crystal, ice, speed.

What Type of Drug It Is: Amphetamine.

What It Does: Meth is a stimulant. A strong one. It elevates mood and reduces appetite. At high doses, it can cause psychosis, seizures, and bleeding in the brain. Chronic use can lead to mood swings and violent behavior, and its use is associated with poor diet and hygiene which can lead to other issues, like severe tooth decay.

How It Does This: Meth activates TAAR1 receptors in the brain, which triggers the release of neurotransmitters like norepinephrine and dopamine at high levels, as well as other changes in neuronal signaling. This has the combined effect of increased wakefulness and activity, rapid heartbeat and breathing, and feelings of extreme euphoria.

What the Risks Are: There are actually medically approved varieties of methamphetamine on the market to treat ADHD and obesity. That said, it's highly addictive, and can kill your brain cells or lead to death. Do not mess around with meth unless you've got a doctor's prescription in hand.

WHAT DO YOUR DREAMS MEAN?

Dreaming is pretty strange. You close your eyes, slip into your subconscious, and watch stories play out in your mind. Some of them are pretty normal—like dreaming about a boring day at work. And some of them can be pretty out there—like dreaming about a dog dressed in a suit who's canoeing with you down a river of lava so you can vote for prom king. So, it's no wonder many of us rouse from our slumber and think, "Why the heck did I dream about *that*?"

Dreams Are a Genie in a Bottle

Psychology has always had its fingers in the dream world, and it started with good ol' Sigmund Freud. (You knew he was going to show up again, right?) Freud believed that dreams reveal the unconscious and that studying dreams could help him understand a person's mind. In fact, he argued that dreams simply provide an outlet for people to carry out their repressed desires. Freud called it "wish fulfillment," a term he actually coined. But of course, this theory was debunked, even during Freud's lifetime, when it became clear that his theory failed to explain clients with PTSD who experienced repeated nightmares.

A Symbolic Gesture

Freud's successor, Carl Jung, believed that dreams are important messages from the unconscious mind—and therefore require very close attention. Through his examination of dreams, Jung identified recurring *archetypes*, which he described as innate, universal symbols for ideas that are part of humanity's collective unconscious. This theory led to Jungian dream analysis, which involves studying a dream, identifying the archetypes, determining their meaning, and applying that meaning to a person's life. Dream analysis became incredibly popular, and there are plenty of websites that can break down dream symbology. Nevertheless,

modern theories have moved away from Jung's belief that every dream has serious significance.

So, What Do They Really Mean?

Although Freud's theory is debunked and Jung's approach is falling out of favor, that doesn't mean that dreams are insignificant. On the contrary, what you dream about probably reflects what your brain considers important. If you feel anxious about taking a test, you may dream about a test (and you forgot to study and you don't know what room it's being held in, and of course you're not wearing any pants) or about something else that makes you anxious. If you spend all day watching *The Lord of the Rings*, you might dream about Gandalf's magnificent beard. And some dreams may provide you meaningful insight into your emotional state. But most dreams are simply regurgitations of experiences from your day—like a DJ sampled your life and was like, "Remix!" Dream interpretation can be really fun, and sometimes therapeutic, but be careful — it can veer into horoscope-level nonsense pretty easily.

I'M DREAMING?

In 1898, a man named Frederik van Eeden wrote down his latest dream. He'd recorded hundreds, but this one was different. In it, he became conscious enough to realize that he was dreaming and could "act voluntarily" within the dream. He called it a "lucid dream." Lucid dreaming occurs most often when we are between REM sleep and being awake. It's believed that certain brain areas—like the dorsolateral prefrontal cortex, which is responsible for working memory, and the parietal lobe, which is responsible for a lot of our sensory information—may become active, giving you conscious control while asleep. If you can harness it, lucid dreaming can improve problem-solving skills, improve self-confidence, and even help combat nightmares. If you'd like to lucid dream yourself, spend some time thinking about what you want to dream about before falling asleep, and pay attention to the minutiae of life so you can better spot breaks from reality in the dream state. Once you're in, the possibilities are endless—you could fly to space, conjure a 100-scoop ice cream cone, or fold the world in half like in *Inception*.

THE FAULT IN OUR MEMORIES (AGAIN)

Now, let's examine the phenomenon of déjà vu: that strange feeling we sometimes get that we've lived through something before. (Hey, wait a second—didn't I just . . . did I have a dream about this?) Dreams and memories have a strange relationship. People who regularly recall their dreams report having a higher number of déjà vu experiences, which may be because dreams promote the creation of false memories. That sounds preposterous, but it's true—dreaming helps us consolidate and simplify memories, which can lead to incorrect information being added based on what feels most familiar to us. This can then make future experiences feel strangely recognizable. In fact, some people report experiencing déjà rêvé, which is the feeling you've *dreamed* something before it has happened in real life. Creepy. Moral of the story: Dreams are not to be trusted.

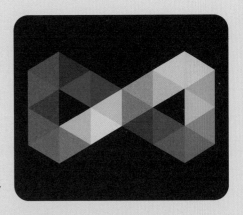

NOT-SO-SWEET DREAMS

Many of us pretty much constantly find ourselves wishing we could get more sleep; for many folks, with work, family, and daily activities, it can be tough to feel like there's time to get in those full eight hours. But we really do need to get that sleep—most adults need between seven and nine hours of sleep every day, and not getting enough sleep is linked to all kinds of long-term health problems, like obesity, high blood pressure, and even dementia. For one reason or another, not everyone can get that full night's rest every night. Here are some of the more common examples of sleep disorders—as well as some of the least common!

Restless Legs Syndrome: As the name implies, people with restless legs syndrome, or Willis-Ekbom disease, have a hard time sleeping because they feel discomfort in their legs when sitting or lying down. They feel the urge to move their legs to alleviate the pain and, in turn, can't get enough sleep. Often, due to the kicking, neither can their partners!

Sleep Apnea: Obstructive sleep apnea is a relatively common condition caused by an obstructed airway, leading to an individual waking up repeatedly throughout the night, gasping for air. It can be caused by physiological traits like a large tongue, oversized tonsils, or obesity. Central sleep apnea is similar, but instead of being caused by the body, it's caused by the brain. Basically, the brain stops telling the breathing muscles to keep doing their thing, causing an individual to choke.

Narcolepsy: Known as a hypersomnolence disorder, meaning excessive sleepiness despite an adequate amount of sleep, narcolepsy is characterized by an irresistible urge to sleep at inappropriate times. It's linked to a lack of neurons that produce hypocretin, which is critical for telling the brain to "wake up" and stay awake.

Shift Work Sleep Disorder (and Other Sleep-Wake Disorders): These are often considered circadian rhythm sleep disorders, because they're generally caused by a disconnect between the body's biological clock and the light/dark cycle of the environment. People who work overnight shifts often end up training their circadian rhythm to be awake at night and asleep during the day, essentially leaving them feeling jet lagged anytime they do have to be out and about during the daytime.

Parasomnias (REM Sleep Behavior Disorder, Exploding Head Syndrome): Sleep disorders that affect the transitional stages between sleep and waking, and between different sleep stages, are called parasomnias. These can include relatively ordinary weirdness like sleepwalking and night terrors, but can also include some extremely strange conditions. One example is REM sleep behavior disorder, where individuals physically and vocally act out their dreams (sometimes injuring themselves or their partner). Another even more bizarre example is exploding head syndrome, where people will "hear" loud explosions in their head as they wake up from sleeping.

Insomnia: The inability to fall or stay asleep, even when you need rest, can be caused by stimulants, such as caffeine, or health conditions like chronic pain, depression, or anxiety. The extremely rare genetic disease, fatal familial insomnia, first identified in an Italian family, is a condition that kills brain cells in the thalamus, which regulates sleep. Victims suffer mild insomnia in middle age, which worsens over a few months. The insomnia leads to hallucinations, confusion, and difficulties with memory, vision, speech, and movement. The disease is eventually fatal. We're not kidding when we say you need to sleep!

YOUR BRAIN ON ...

GHB

What It Is: Gamma-hydroxybutyric acid, aka GHB, Liquid Ecstasy, G, or most frighteningly, "the date rape drug."

What Type of Drug It Is: Sedative.

What It Does: GHB is used to treat sleep disorders like narcolepsy (usually prescribed opposite a low-dose daytime stimulant, to help restore normal sleep habits), due to its strong sedative effects. It also can induce feelings of euphoria and disinhibition, making it popular among some recreational users.

How It Does This: GHB the drug is actually a precursor to other neurotransmitters, including GABA, which is an inhibitory transmitter in the brain. GHB itself can bind to the GHB and GABA$_B$ receptors, which sets off cascades of signals that lead to the sedative and euphoric effects.

What the Risks Are: GHB has a depressing effect on the central nervous system, similar to alcohol, and it can be lethal at high doses. Some athletes use GHB because it's marketed as an "anabolic agent," but there's no evidence that it improves performance. And of course, as you may have learned in drug-awareness or personal safety classes, GHB has a reputation as a date rape drug, due to several high-profile criminal cases.

SLEEP MAKES
SQUEAKY-CLEAN BRAINS

There's a growing body of evidence that poor sleep habits play a role in the development of dementia, and people with dementia frequently experience serious sleep disruptions. We're not totally sure why, but recent research in mice has given us some clues. During sleep, brain cells called astrocytes contract, expanding the space between all the brain cells by almost 60 percent.

This allows more cerebrospinal fluid, or CSF, to flow through the cracks in the brain. When this happens, the CSF acts like a cleaning fluid, pushing out all of the debris that collects during the day. Clearing out these toxic waste products, like amyloid beta protein, may be important for preventing diseases such as Alzheimer's.

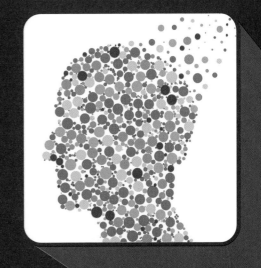

Chapter 14

WHAT IS LOVE?
(BABY, DON'T HURT ME)

If you believe the movies, love is like oxygen. Love is a many-splendored thing . . . love lifts us up where we belong . . . all you need is love . . . or is it just your brain trying to make sure you get it on and make some babies?

Humans are obsessed with love. We write books, songs, and movies about it, we need it from our parents, we want it from our friends, and for many, finding true romantic love is one of life's primary goals. But love isn't a tangible need like food or water or even sleep; it's an abstract idea that reflects our relationships with the people around us. So—what is love? And why do we crave it so badly?

Human beings are deeply social creatures. Our big, beautiful brains are a direct result of our early ancestors' willingness to form cooperative communities, where multiple adults could help care for and protect weak, helpless infants, forage for shared food, and coordinate together on hunts. Being able to form strong bonds with one another kept our early ancestors alive, and it's

what keeps us from murdering our siblings today.

Love probably most often conjures thoughts of red roses, champagne, and—*cough*—well, you know. Humans are animals, too, and we've gotta make sure the next generation gets, uhh . . . generated. Romantic love may have evolved to ensure that both DNA-providing parents stuck around long enough to see their offspring make it out on their own.

But love also comes in other forms. The love a parent feels for their child ensures that the parent keeps the kid alive, even through those 3 a.m. screaming fits. The love you feel for your friends can get you through difficult exams and hard days at work. And all those ooey-gooey, lovey-dovey feelings come from—you guessed it—that silly squishy brain of yours!

THIS IS YOUR BRAIN ON LOVE

Falling in love is a dizzying, terrifying, exhilarating experience—one that affects us emotionally *and* physically. Ever feel those butterflies in your stomach? Or get that rush to your head when your crush brushes your hand? There's a lot going on up in that noggin of yours when you're falling in love.

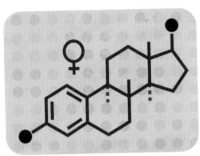

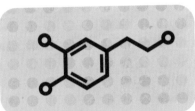

Dopamine: Contributing to the euphoric sensation of being in love, dopamine activates a lot of the reward circuitry in the brain. When writers claim that "love is just like a drug," what they really mean is that falling in love activates a lot of the same pathways as using stimulant drugs, such as amphetamines and cocaine.

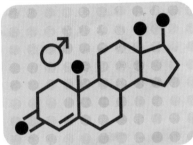

Sex Hormones: Those traditional sex hormones, estrogen and testosterone, are part of what drives that irresistible lust feeling. Interestingly, testosterone goes down in males but goes up in females during this passionate phase.

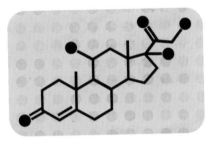

Cortisol: At the start of a new romantic relationship, levels of cortisol—a stress hormone—quickly rise, which contributes to the excited, nervous butterflies you might feel in your stomach.

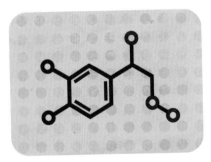

Norepinephrine: Also known as adrenaline, norepinephrine is a neurotransmitter that gets you up and at 'em, and it's part of why you feel so giddy and excited when you're crushing hard.

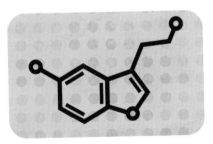

Serotonin: Unlike dopamine and norepinephrine, levels of serotonin go down when we're infatuated. Low serotonin is actually a marker for obsessive compulsive disorder, and researchers think this is why you can't get your crush out of your head!

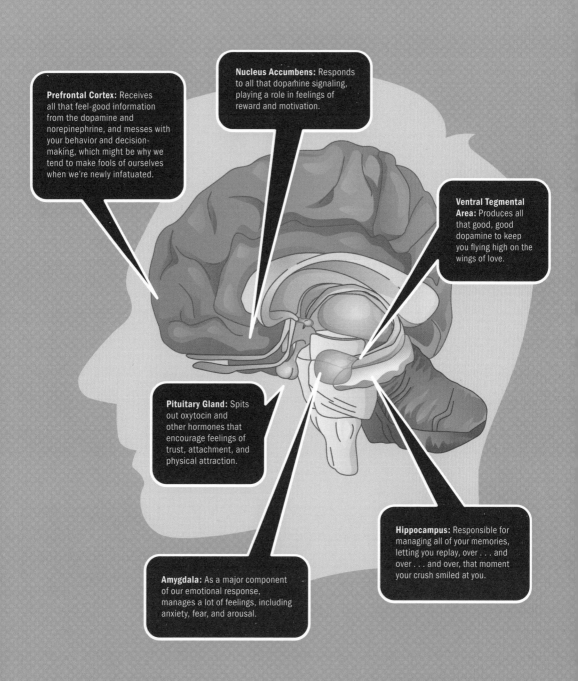

Prefrontal Cortex: Receives all that feel-good information from the dopamine and norepinephrine, and messes with your behavior and decision-making, which might be why we tend to make fools of ourselves when we're newly infatuated.

Nucleus Accumbens: Responds to all that dopamine signaling, playing a role in feelings of reward and motivation.

Ventral Tegmental Area: Produces all that good, good dopamine to keep you flying high on the wings of love.

Pituitary Gland: Spits out oxytocin and other hormones that encourage feelings of trust, attachment, and physical attraction.

Amygdala: As a major component of our emotional response, manages a lot of feelings, including anxiety, fear, and arousal.

Hippocampus: Responsible for managing all of your memories, letting you replay, over . . . and over . . . and over, that moment your crush smiled at you.

LOVE FROM EVERY ANGLE

The Greeks had six words to specify various kinds of love: *agápe*, *philía*, *storgē*, *philautia*, *xenia*, and *éros*. Only one of those words, *éros*, describes sexual love. The rest are reserved for love of God, love of friends, love of family, and even love for houseguests. That's all good and great, but psychology may have a more useful way of understanding love, particularly nonromantic love.

Love Triangle

One model, called the triangular theory, suggests that there are three elements of any kind of love: intimacy, passion, and commitment. Like a recipe, different combinations of these three ingredients results in different flavors of love. If you combine passion and intimacy, that's romantic love. Combine intimacy and commitment and you have compassionate love, like a parent's love for their child. Just have intimacy on its own? That's a friendship. So many combinations!

Blood Is Thicker Than Water

Nonromantic love is one of our most basic and fundamental needs from birth. Our intense desire to form close emotional bonds with others, called attachments (see page 64 for more), are important for development, but also probably for evolution. If you found yourself in a *Vertical Limit*-type situation where you could only save either your sister or your climbing buddy, you're (probably) gonna save your sister. This is called kin selection, and it essentially explains how you give preferential treatment to the people you love and feel closest to (in other words, family), which will lead to some of your genes being passed on through their offspring. But if this is true, then it's interesting that most people don't take that a step further and make babies with their own siblings, right?

Love You Like a Sister

International sensation and official dirty boy Sigmund Freud believed that family members are naturally sexually attracted to one another, which made it necessary to develop incest taboos. But there's little evidence to support Freud's claims. On the flip side, the Westermarck effect hypothesizes that children who are raised in close proximity to one another from a young age show no substantial sexual attraction to one another. When studying collective communities in Israel, children who grew up in the same peer group from birth to age six showed no sexual attraction to one another and never married kids from the same peer group (despite intergroup marriage being quite common).

WHY DO YOU LOVE MR. FLUFFERS?

Our ancestors started taming dogs and other animals about 10,000 years ago. They served a purpose as "living tools"—wolves helped hunt, chickens laid eggs, horses provided transportation, and so on. But since then, pets have taken over our homes. Dogs, cats . . . hedgehogs?! They're cute, but they also require lots of time and money. Why do we keep these little guys around? Well, we're not quite sure. Some argue that keeping pets is a social construct, like fashion—you see other people doing it and think, "Hey, I should do that!" Others think pet ownership used to be a symbol of wealth or of your nurturing nature and, therefore, value in procreating. But what seems most likely is that humans are just really social and we crave close connections and attachments with others, human or otherwise. Like children, having a pet that completely depends on us triggers our protective and nurturing instincts, and validates our own existence.

THE PLATONIC BRAIN

As we explored earlier, the sexual passion associated with falling in love is like dropping a dirty bomb in your brain filled with a mess of neurotransmitters and hormones. But in platonic, nonromantic love, such as the love you have for your friends, your children, your parents, or your cat, the brain is a bit more selective in its hormone selection.

A Touch of Love

If you've ever received a really good hug from a friend or spent a few peaceful minutes slowly stroking your cat's fur or stared knowingly into someone else's eyes, you've likely felt the effects of oxytocin. Oxytocin is sometimes called the trust hormone. It's often released during acts of touching, particularly when there is skin-to-skin contact. Oxytocin makes us feel calm, secure, and comfortable—it essentially helps us feel closer to others. In the body, oxytocin also triggers contractions and lactation during pregnancy and birth, and is vital for mother-infant bonding.

Stuck on You

On the other hand, part of what keeps us continuously caring for our friends, family, and loved ones is a fun hormone called vasopressin. Known primarily for its role in maintaining blood pressure and cardiovascular function, vasopressin is sometimes referred to as the attachment hormone because its release is associated with feelings of affection. Vasopressin also appears to be involved in the more "active" side of love, making us feel possessive, protective, and attuned to the needs of the people that we care about. This hormone gets released with oxytocin, and they work in tandem to help us feel that warm, intimate, exclusive connection with others.

MOMMY'S LITTLE HELPER: OXYTOCIN

When your kid melts down for the fourth time in three hours, screaming on the floor of the cereal aisle, you might find yourself wondering how it is that humans have managed to keep their irrational, difficult children alive for all of these generations. The answer, it turns out, is hormones—and especially oxytocin. In addition to supporting the physical aspects of baby-making (oxytocin stimulates contractions, lactation, and sexual arousal), it's important for initiating parental bonding, ensuring that you will love and protect your terrible toddler at all costs. Research has shown that blocking oxytocin in female rats causes them to neglect their offspring; on the flip side, injecting unbred sheep or rats with oxytocin causes them to respond and provide care to infants that don't belong to them.

LET'S TALK ABOUT SEX, BABY

It's easy to get humans hot and bothered. Sexual arousal, stimulation, and orgasm are all part of the machinery that ensures we keep producing more tiny humans so our species goes on existing. And when you're aroused, it can feel like your brain has completely flown the coop, leaving you to flounder in your sexual desires. But that big ol' brain of yours is still around, twirling the dials on your love-o-meter to get you all frisky.

Start Your Engines

Sexual arousal often originates psychologically, which is why reading a dirty romance novel or watching a steamy movie can help get things revved up when you're trying to get in the mood. But arousal can also arise from physical stimulation of erogenous zones innervated by neurons from nerves in the lower spine, leading to the sensations we associate with being aroused.

It's Getting Hot in Here

Sexual arousal has a lot of physical effects—breathing becomes more rapid, the heart speeds up, and, perhaps most notably, various bits and bobs start to swell with blood in anticipation of intercourse. These bodily functions are controlled by the parasympathetic nervous system, one component of the autonomic nervous system. The parasympathetic nervous system oversees "feed and breed" activities, such as digestion and sex, via nerves originating in the brain stem and lower spinal cord. In the case of sexual arousal, those lower spinal cord nerves receive and send information to and from the genitals, causing increased blood flow to generate erections and lubrication.

Doing the Dirty

During sexual activity, all that erotic sensation builds up and, in some cases, results in an orgasm. Despite being portrayed in media (especially in erotic novels) as some sort of mind-altering, galaxy-bending experience, orgasms are just another experience somewhere on the pleasurable end of the human-sensation spectrum. When sexual sensation reaches a certain threshold, signals from the spinal cord trigger the release of muscular tension and blood flow; in people with penises, this also usually includes ejaculation.

In the brain, a whole lot of stuff is going on during an orgasm. Using fMRI, researchers have seen decreased activity in the amygdala (associated with feelings of fear and anxiety) and the orbitofrontal cortex (associated with impulse control). In contrast, there is increased neuronal activity in areas associated with our sense of touch, emotional control, memory, and decision-making.

All in all, sex is pretty good for your brain. There's even evidence that sexual stimulation and orgasms can decrease pain sensitivity, improve symptoms of anxiety and depression, self-esteem, quality of sleep, and levels of partner intimacy.

HOW TINDER CHANGED DATING

Swipe left—no, right! Technology has fundamentally altered the dating landscape, making it easier than ever to meet someone new. But those dating apps also alter your brain. Every time you match with a hot new single in your area, your brain gets a hit of dopamine and you feel good. But of course, you don't match with every person. The unpredictable nature of the reward results in larger spikes in your pleasure centers, like the nucleus accumbens. As a result of this immediate gratification, some say, these apps work against the "intended" effect of finding long-lasting love because only about 0.2 percent of matches result in someone getting a phone number and even fewer of those virtual interactions translate into actual get-togethers. It just becomes a dating video game. So forget about the rush of going on a first date; matching can be just as exciting!

ADDICTED TO YOUR LOVIN'

Sex and pornography are fairly regular parts of life. They've existed throughout all recorded time and beyond. (There was probably some really great cave-painted porn.) But despite its ubiquity, topics of sexuality also attract controversy—particularly when it comes to addictions. You may be quite familiar with the idea of someone having a sex addiction or porn addiction. It's gossiped about with friends or discussed in films. And with the ubiquity of the internet, these addictions seem more likely than ever thanks to dating apps and porn sites. Sexy times for everyone!

Just Can't Get Enough

But simply enjoying porn or having frequent sex doesn't make you addicted to either thing. So, when do you have a problem? In practice, problems might involve having obsessive sexual thoughts, spending a large amount of time seeking out sex or watching porn, or engaging in compulsive encounters with multiple partners. Importantly, for this to even come close to a "disorder," the sufferer will likely have trouble controlling their behaviors, feel guilty for their sexual behavior, and lie to cover up their

behaviors. They may become so preoccupied with sex or porn that it interferes with their daily life, even resulting in real-life bad stuff happening to them professionally or personally. Yikes—sounds serious!

Compulsion, Not Addiction

And yet, despite "porn addiction" and "sex addiction" being common colloquial terms, neither condition has made its way into the DSM or ICD as formal disorders. Instead, there are alternative diagnoses with markedly unsexy names like "Other Sexual Dysfunction Not Due to a Substance or Known Physiological Condition." For a while, experts considered a new label, hypersexual disorder, which would cover both sex addiction and porn addiction. They ultimately decided to forgo this diagnosis because the body of evidence was pretty slim and, in particular, because sex and porn do not evoke the same biological changes as, say, drugs or gambling that would prompt anyone to classify it as an addiction. For now, it's been deemed a mere compulsion, a repetitive behavior with no rational motivation except perhaps the intent to reduce anxiety.

WHAT WE GOT WRONG ABOUT SEX

Sex is often surrounded with mystery. But don't believe the things you learned at recess. These are all just myths.

"Porn kills your relationship satisfaction."

Although it can be a problem if one person sees porn as bad, most people in relationships watch porn and report being attracted to their partner and having higher relationship satisfaction overall.

"Masturbation causes mental health problems."

Nope! If you're raised hearing it's perverted or bad, then it might cause guilt. But otherwise, it's normal and healthy and doesn't cause mental health problems.

"Men can't be raped."

Of course they can. About 3 percent of American men have been victims of sexual assault at the hands of both men and women—in fact, 10 percent of all rape victims are male.

"Sex decreases your athletic performance."

Wrong again. It might feel intuitive that a demanding activity like sex will make you feel weaker, but there's no evidence to support that old coach's tale.

WHAT'S GENDER GOT TO DO WITH IT?

It seems like we can't go more than a couple of years before some out-of-touch academic says something like, "Women are bad at science because that's just the way their brains are built—DUH," or "Men are more rational—that's why they're so good at business!" These claims are frankly insulting, but where do they come from? And are there any actual differences between the male and female brains?

Men Are from Mars

Famously, fMRI studies looking at brain structures and connectivity in male and female brains have found striking differences; namely that men have six and a half times as much gray matter compared to women, indicating a superior ability for logical decision-making and emotional control. In contrast, women have ten times as much white matter as men, which of course means that their brains are more interconnected and better suited for integrating information and multitasking. It seems pretty convenient that these neuroimaging studies would line up with existing stereotypes about men and women! Of course, it's never that easy.

Women Are from Venus

As many scientists have pointed out, once you account for differences in brain size (on average, men's brains are slightly larger than women's), nearly all of these "obvious" structural differences disappear. There are a few "real" differences between male and female brains, when you look at averages among genders. For example, some studies have found that men tend to have more gray matter on the left side of the brain, while women have more on the right. Women tend to have higher levels of estrogen, which has effects on cognition. And on average, men perform a little better on spatial tasks (like solving a maze), while women perform a little better on verbal memory tasks (remembering words). But these are just averages—the truth is that the spread of variation among members of the same sex is way higher than the spread of the differences between sexes. Plus, none of this information says anything meaningful about a person's emotional or cognitive abilities.

But What About the Rest of Us?

All of this doesn't even address individuals who don't identify with a binary gender, and most studies on gender in the brain don't include transgender individuals. Some studies have found slight differences in brain structures like the amygdala and hypothalamus in transgender women compared to cisgender men, but, again, that doesn't necessarily mean anything about a person's abilities or behavior.

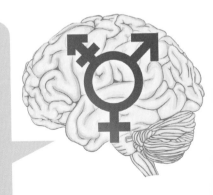

THE TRANSGENDER BRAIN

As much as some people would love to be able to easily categorize genders as penis = male, vagina = female, biology is never so simple. Like we said above, there actually aren't that many differences between male and female brains, and the variation within a gender is generally much higher than the variation across genders. That said, recent research on brain-activity patterns has found that those of trans people tend to look more like the activity patterns of the gender they identify with, as opposed to those of the gender they were assigned at birth. We're not exactly sure why some folks are transgender, but studies have indicated some underlying biological effects, possibly related to prenatal exposure to sexual hormones. And this isn't even touching on people who are intersex and/or nonbinary/gender-nonconforming!

If you think gender in the brain gets complicated, wait until we start talking about sexuality. You love who you love—but why?

It's in the Gene Pool

While there's no single "gay gene," a recent enormous study of the human genome and its relationship to human sexuality found a number of genetic markers with weak associations to homosexual behavior. Some of these markers are found close to or are associated with genes controlling things like testosterone levels and the sense of smell, which have been linked to sexual attraction in the past. But all in all, these genetic markers can account for only maybe 15% of nonheterosexual behavior—the rest is likely environmental.

You Can Thank Your Mother for That

There's a bunch of research out there looking at the ways in which environment might affect sexual orientation, mostly about how certain exposures during pregnancy can affect sexuality in cisgender men. Some research indicates that variations in sex hormone levels (like testosterone and estrogen) in the womb can have an effect on sexual orientation. Weirdly, if you're male, your place in your family's birth-order and the number of older brothers you have can also influence your chances of identifying as homosexual. Called the fraternal birth order effect, it basically boils down to this: The more older brothers you have, the more likely you are to be gay. This is thought to be a result of how a woman's immune system responds to the Y-chromosome-linked proteins produced by a male fetus, which can in turn influence neurodevelopment in future male infants. And, strangely, this effect is not seen in gay women—or in left-handed men!

Sexuality in the Brain

Interestingly, some neuroimaging studies have found sexually dimorphic (meaning different between male and female) differences in a particular region of the hypothalamus, known as the sexually dimorphic nucleus of the preoptic area. (Whew—what a mouthful.) It's thought that this brain region assists with processing sexual stimuli, and there are differences in size and cell number between hetero- and homosexual men, as well as differences in activation patterns in hetero- and homosexual men and women.

Hey, We're Here Too!

Like we said before, a lot of the research on this topic has focused on sexuality in cisgender men (the patriarchy, amirite?), so we don't know too much about other kinds of sexual orientations, like bisexuality and pansexuality, or asexual or aromantic orientations. In fact, some circles still consider some of these sexualities to be a myth, or even a disorder. But with growing social awareness of the reality of these orientations, some scientists are trying to better understand what, if anything, is going on biologically in people who identify with them. One study, for example, found that asexual men are more likely to be a younger sibling, while asexual women are more likely to be an older sibling.

SEXISM AND THE SINGLE BRAIN

Is it possible that there are real, meaningful differences between the brains of different genders? Sure. Is it also likely that most research on this subject is based on a long history of sexism and a desire to "find an answer" to problems that are probably caused by socialization and cultural expectations rather than biology? Also yes! Most of this research sets out to spot the differences, and the way a question is asked can bias the results of the research—if you go looking for evidence that male and female brains are different, and that those differences mean that men are smarter and women are more emotional, you're pretty likely to find something that supports your hypothesis.

THE MAN WHO CHANGED HOW WE THINK ABOUT SEX

Part of why the neurobiology of human sexuality is so poorly understood is because for most of the history of white Western society, identifying as anything other than a heterosexual cisgender person was totally taboo. The general attitude was that feeling anything else made you a sexual deviant, possibly dangerous, and, at the very least, damaged. But even within these repressive boundaries, some folks have pushed back. One of the most well-known individuals in this space is Alfred Kinsey, a bisexual man who made it his life's mission to better understand human sexuality.

A Renaissance Man

Kinsey originally trained as an entomologist and as a young man spent his time studying wasps. Over time, he got interested in their variations in mating habits and began to discuss sexual behavior and practices with his colleagues, eventually leading a course on marriage and sexuality at the university where he taught. He even developed a scale to measure sexual orientation, known as the Kinsey scale (more on that in a minute).

A Sexpert, Indeed

Kinsey is seen by many as the first major figure in American sexology—the study of human sexuality. He was a huge proponent of sexual freedom, loudly denouncing repressive laws and social norms. He argued that rather than being sexually deviant, most so-called "perversions" actually fell somewhere within the normal range of human sexuality. His interest in sexuality led him to conduct thousands of interviews, collecting information on people's sexual histories, and he published them with the help of his wife, Clara McMillen. These books provided new insights into the prevalence of behaviors like masturbation and homosexuality, and showed that despite what social norms had to say about it, all kinds of sexual behaviors and relationships are enjoyed by individuals of all our various genders.

. . . and an Imperfect Hero

Like so many folks, Kinsey did a lot of great things for helping liberate human sexuality in American society, but he also did some things that were pretty questionable, like actively engage in sexual activity with some of his research subjects (he later argued this was to help gain the trust of said subjects) and even his coworkers. He also fudged a lot of details about information he included in his studies, and some argue that his interviewees were not very representative of the general population—too many prostitutes, not enough housewives, that kind of thing. Because of these critiques, his conclusions should be taken with a grain of salt, but it doesn't change the fact that his work gave us a pretty big push toward greater acceptance of a wide range of sexual behaviors.

HORMONE MYTHBUSTING!

The male sex hormone testosterone has been tied to all kinds of "masculine" behaviors, like dominance, aggression, and violence. Biologically speaking, testosterone is important for male sexual maturation, such as deepening the voice and promoting facial-hair growth. It's also a factor in muscle size and strength and sex drive. But women have testosterone, too! In females, testosterone plays a role in ovarian function, bone strength, and sex drive. As with so many things, testosterone tends to be best in moderation; while higher testosterone levels in males have been linked to higher mating success and increased perceived attractiveness, too much testosterone causes loads of problems, like low sperm count, weight gain, mood changes, and increased risk of heart damage and liver disease.

WHERE DO YOU FALL? ▶ THE KINSEY SCALE

One of the most memorable things Alfred Kinsey gave us is the Kinsey scale, also called the Heterosexual-Homosexual Rating Scale. This scale is used in research settings (and sometimes, when a few drinks have been consumed, in social ones) to describe an individual's sexual orientation. On the scale, a 0 means "exclusively heterosexual," while a 6 means "exclusively homosexual." Kinsey also used X to denote "no sexual contacts"—an old-fashioned way of saying "asexual." Such a scale helps tear down the idea that sexuality fits into strict, binary categories, and provides a more fluid way of thinking about things. As a female, saying you're a 1 on the Kinsey scale is like an easier way of saying, "Well, I *mostly* like men, but I *also* like Gillian Anderson in a white tuxedo."

0	1	2	3	4	5	6
EXCLUSIVELY HETEROSEXUAL	MOSTLY HETEROSEXUAL SLIGHTLY HOMOSEXUAL	MOSTLY HETEROSEXUAL MORE THAN SLIGHTLY HOMOSEXUAL	EQUALLY HETEROSEXUAL AND HOMOSEXUAL	MOSTLY HOMOSEXUAL MORE THAN SLIGHTLY HETEROSEXUAL	MOSTLY HOMOSEXUAL SLIGHTLY HETEROSEXUAL	EXCLUSIVELY HOMOSEXUAL

IT'S JUST HORMONES

You know that stereotype about how when people are menstruating, suddenly they can't stand their partners? That's probably mostly a myth, like so many things that make hormonal women out to be total monsters, but there is some evidence that your hormonal status can affect who you're attracted to.

Aunt Flow Has Some Opinions

There's some research that who you find attractive varies based on where you are in your menstrual cycle; the theory is that when you're close to ovulating, you get the hots for folks who are the most genetically fit to help you produce the healthiest baby, while at other points in the cycle, you're more interested in a partner whose traits can provide stability and support for raising a child. While some studies have found that people close to ovulation are more interested in having sex outside of their relationship, it doesn't necessarily mean that it

has anything to do with evolutionary fitness, and might actually have more to do with a person's mood and how they interact with their partner than anything else. So, folks, maybe think about bringing your partner some chocolate even when they're not on their period.

Pills, Chills, and Thrills

There's also the question of what, exactly, makes a person sexy, and whether that changes based on a person's hormone status. Some research in cisgender people has suggested that women who are on hormonal birth control when they begin a relationship are more likely to be paired with a comparatively feminine man, while women who weren't using hormonal birth control at the start of their relationship tend to be with dudes who are more . . . macho. More recent studies, however, have not found the same results, nor have they found evidence that your fertility status affects how masculine you like your men.

Chapter 15

SAD, SCARED, AND SELF-MEDICATED

We all have our ups and downs. But when our brains fall out of whack on a microscopic level, that can really mess with our emotions, which have a huge impact on our mental well-being. And, boy, folks have found ways to cope with that, too.

You've probably felt anxious before in the face of stressful events. And surely your mood changes when bad things happen. Yet for some folks, these feelings become so severe and overwhelming that they interfere with job responsibilities, personal relationships, and even daily functioning. Depression, anxiety, and other mood disorders occupy top spots among the most common mental disorders in the world and affect millions of people.

It's a widely held myth that depression isn't a real disorder at all. It's falsely believed that depression and sadness are the same thing, that people who experience depression are just lazy or mentally weak. But, depression isn't just the result of having a tough time. Rather, it's a serious medical illness that's marked by symptoms like loss of interest or pleasure in activities, depressed mood, changes in sleeping and eating habits, loss of energy, difficulty concentrating or thinking, and suicidal thoughts.

Misconceptions have plagued anxiety, too.

But people who suffer from anxiety aren't just worrywarts or neurotic. No, anxiety disorders differ from normal feelings of stress. Folks with these disorders experience fear and anxiety at a level that is out of proportion to the situation and to the point that it affects their normal functioning. Symptoms vary, but can include panic attacks, social withdrawal, avoidance of triggers, difficulty relaxing, racing thoughts, irritability, and much more.

And finally, a great number of untrue stereotypes exist about other mood disorders. Although we often think of these disorders as distinct issues, the truth is that they are sometimes two sides of the same coin. Depression and anxiety often go hand in hand and provoke each other, causing a vicious cycle where one feeds the other. These conditions are often linked to other mood disorders, too, like bipolar disorder and PTSD. It appears that just as our brain's wiring is interconnected, so too are its dysfunctions.

THE BLUES IN THE BRAIN

Your brain is responsible for producing all kinds of emotions—including the negative ones. Emotions are complicated and can't be easily pinned down, but there are some brain regions that appear to play particularly large roles in things like depression, anxiety, and fear. In particular, the limbic system is important for emotional regulation—so, when the limbic system isn't happy, ain't nobody happy.

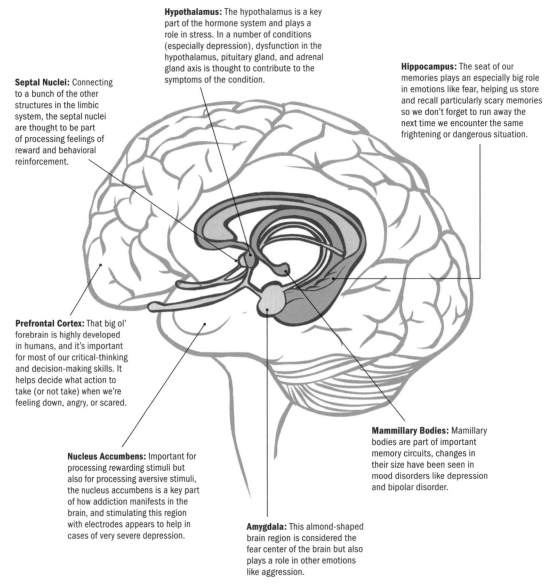

Hypothalamus: The hypothalamus is a key part of the hormone system and plays a role in stress. In a number of conditions (especially depression), dysfunction in the hypothalamus, pituitary gland, and adrenal gland axis is thought to contribute to the symptoms of the condition.

Hippocampus: The seat of our memories plays an especially big role in emotions like fear, helping us store and recall particularly scary memories so we don't forget to run away the next time we encounter the same frightening or dangerous situation.

Septal Nuclei: Connecting to a bunch of the other structures in the limbic system, the septal nuclei are thought to be part of processing feelings of reward and behavioral reinforcement.

Prefrontal Cortex: That big ol' forebrain is highly developed in humans, and it's important for most of our critical-thinking and decision-making skills. It helps decide what action to take (or not take) when we're feeling down, angry, or scared.

Mammillary Bodies: Mamillary bodies are part of important memory circuits, changes in their size have been seen in mood disorders like depression and bipolar disorder.

Nucleus Accumbens: Important for processing rewarding stimuli but also for processing aversive stimuli, the nucleus accumbens is a key part of how addiction manifests in the brain, and stimulating this region with electrodes appears to help in cases of very severe depression.

Amygdala: This almond-shaped brain region is considered the fear center of the brain but also plays a role in other emotions like aggression.

MEDS ON THE MOLECULAR LEVEL

Neurons exchange information by exchanging chemical messengers at connection points called synapses; these chemicals, called neurotransmitters, keep the brain's electrical signals flowing between cells. In the case of conditions like depression and anxiety, there's some evidence that medications affecting neurotransmitter release and binding can help alleviate some of the symptoms of mental illness. Here's how SSRIs, MAOIs, and SNRIs all work.

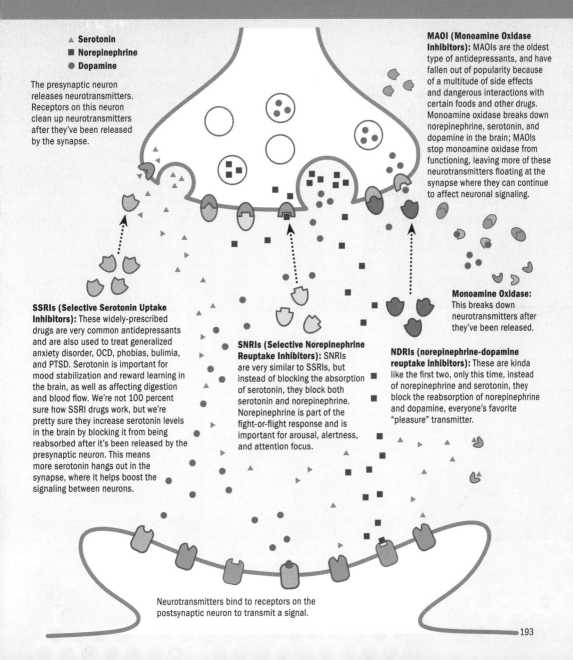

▲ Serotonin
■ Norepinephrine
● Dopamine

The presynaptic neuron releases neurotransmitters. Receptors on this neuron clean up neurotransmitters after they've been released by the synapse.

MAOI (Monoamine Oxidase Inhibitors): MAOIs are the oldest type of antidepressants, and have fallen out of popularity because of a multitude of side effects and dangerous interactions with certain foods and other drugs. Monoamine oxidase breaks down norepinephrine, serotonin, and dopamine in the brain; MAOIs stop monoamine oxidase from functioning, leaving more of these neurotransmitters floating at the synapse where they can continue to affect neuronal signaling.

SSRIs (Selective Serotonin Uptake Inhibitors): These widely-prescribed drugs are very common antidepressants and are also used to treat generalized anxiety disorder, OCD, phobias, bulimia, and PTSD. Serotonin is important for mood stabilization and reward learning in the brain, as well as affecting digestion and blood flow. We're not 100 percent sure how SSRI drugs work, but we're pretty sure they increase serotonin levels in the brain by blocking it from being reabsorbed after it's been released by the presynaptic neuron. This means more serotonin hangs out in the synapse, where it helps boost the signaling between neurons.

SNRIs (Selective Norepinephrine Reuptake Inhibitors): SNRIs are very similar to SSRIs, but instead of blocking the absorption of serotonin, they block both serotonin and norepinephrine. Norepinephrine is part of the fight-or-flight response and is important for arousal, alertness, and attention focus.

Monoamine Oxidase: This breaks down neurotransmitters after they've been released.

NDRIs (norepinephrine-dopamine reuptake inhibitors): These are kinda like the first two, only this time, instead of norepinephrine and serotonin, they block the reabsorption of norepinephrine and dopamine, everyone's favorite "pleasure" transmitter.

Neurotransmitters bind to receptors on the postsynaptic neuron to transmit a signal.

193

DEPRESSING DETAILS ABOUT DEPRESSION

Everyone feels down in the dumps sometimes, but sometimes that sadness you're feeling isn't just the blues. When you're feeling symptoms like sadness, emptiness, hopelessness, fatigue, irritability, changes in eating and sleeping patterns, and even thoughts of suicide, and those feelings are interfering with your life for two weeks or more, you may be diagnosed with major depressive disorder. If you're feeling this way, you're not alone—the National Institutes for Mental Health estimate that nearly 10 percent of the population may experience at least one major depressive episode in any given year. When there's a lot of upheaval in the world—like, oh, say, a global pandemic—that number might go up.

It May Be All in Your Head, But It's Still Real

Because depression can look like normal sadness and it doesn't always have a clear cause, sometimes it's hard for people to believe that it's real. That may be why some people argue that if you just try hard enough, you can break the cycle—you're in control of your own emotions, so you can choose to feel better! But anyone who's suffered from depression knows that that's definitely not true, and the science backs it up; depression is a disease, not a choice. There's definitely a genetic component to depression. If your parent or sibling has the condition, it's more likely that you might end up with it, too. But it can also be linked to environmental factors—things like abuse, neglect, and severe life stressors like financial instability or a death in the family all make it more likely that a person might develop depression.

Subpar Serotonin Signals Symptoms of Sadness

Until recently, scientists were pretty sure that depression is caused by an imbalance of certain chemicals within the brain. Dubbed the monoamine hypothesis of depression, this theory said that a lack of neurotransmitters like serotonin and dopamine—two chemicals associated with feelings of pleasure and reward—would lead to depression. This is why antidepressant medications like selective serotonin reuptake inhibitors, or SSRIs, are considered a first-line treatment for the condition. These medications function by altering the release of serotonin at the synapse, adjusting serotonin signaling throughout the brain.

But Wait—There's More

Scientists are no longer positive that depression is a direct result of a simple chemical imbalance; some of the research and treatments have been inconsistent with that idea, and the condition is really complex. So, researchers are also exploring a lot of other possible explanations.

More recent evidence implicates a number of potential changes in brain structure and function; patients with depression generally have less gray matter in several brain regions, like the cingulate cortex, the hippocampus, and the amygdala, leading to the hypothesis that depression shrinks the parts of the brain that are important for controlling our emotions and making choices to take care of ourselves, making it hard for patients to recover from feelings of sadness.

Research has also found that patients with depression have unusually high levels of cytokines, small proteins that play a big role in the inflammatory response that kicks off when your body is responding to injury. Cytokines are involved in sickness behavior—you know how, when you're sick, you're usually not very hungry, and have trouble concentrating, and really just want to sleep a whole bunch? Hey . . . those symptoms sound an awful lot like depression, don't they? This theory proposes that high cytokine levels, caused by stress, can impact normal neurotransmitter signaling, leading to the symptoms of depression. This is why some doctors may encourage their patients to take anti-inflammatory medications like ibuprofen alongside an antidepressant.

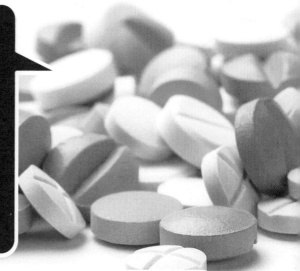

DO PSYCH DRUGS WORK?

You may have heard that antidepressant medications, and especially SSRIs, don't really work that well. It's true that in clinical trials, if you think you're getting an antidepressant, lots of patients end up feeling like their symptoms improved even if they were taking a placebo (basically, a sugar pill with no real medical effects). So, does that mean you shouldn't be taking them at all? The consensus these days is that antidepressants may be helpful in some situations and not in others; it really depends on the person. Our brains are basically big chemical soups, and it's not easy to predict whether a medication that works will for one person will work well for others—plus, the placebo effect can be absurdly strong! Ultimately, it comes down to the individual and their medical team to decide what the best treatments are for them.

WHEN NOTHING ELSE WORKS

While many folks experience depression at different times in their lives and recover with time, therapy, and/or medication, sometimes depression just won't quit. This is known as treatment-resistant depression, and it's definitely hard to treat. Some options include trying nontraditional medications, like mood stabilizers or antipsychotics, to see whether those have an effect. You might also ask about having genetic testing done to see whether you have a specific genetic variation that makes it tough for your body to process certain medications. When all else fails, doctors may consider treatments like electroconvulsive therapy, or ECT—essentially, inducing a small seizure in your brain while you sleep. We're not sure why it works, but repeated ECT treatments do seem to significantly alleviate depression symptoms for lots of folks. Other treatments being tested include ketamine, a noted party drug—more on that later.

MANIC AT THE DISCO

While bipolar disorder often gets lumped in with depression (and in fact this particular diagnosis used to be called *manic depression*), now we know that it's really a very different condition. It can vary in severity and time course, but in its most classic form, an individual will experience episodes at two opposite ends of the emotional scale. On one end are manic episodes, where the person is extremely energetic and may feel euphoric, agitated, and distracted, and make poor decisions; sometimes mania is so severe that a person experiences psychosis and requires immediate medical treatment. On the other end are depressive episodes, which resemble depression.

Where Is My Mind?

As with depression, it's still not totally clear exactly what causes bipolar disorder, but it does seem to have some genetic links—if you have a twin with bipolar disorder, there's about a 60 to 80 percent chance you will, too, and there are some known genetic markers that have been associated with a higher risk of developing the condition. There are also environmental risk factors, such as drug abuse and exposure to severe trauma.

Shifts in the Structure

One of the theories of bipolar disorder is that it's related to structural differences in brain regions involved in cognitive tasks and emotional processing; there's evidence that some regions of the cortex, like the anterior cingulate cortex and ventral prefrontal cortex, are smaller than usual in people with bipolar disorder, while other areas, like the globus pallidus and amygdala, are bigger. It also seems like people with bipolar disorder might have changes in their white matter, which is involved in sending signals between different regions of the brain.

Altered Emotional Circuits

When researchers look at brain-activity patterns in various casess, they find that people with bipolar disorder show less activity in the ventral prefrontal cortex, which plays a big role in emotional regulation; this might mean that without adequate signaling from the ventral prefrontal cortex, brain regions like the amygdala can get hyperactive and contribute to the symptoms of the condition. Interestingly, manic episodes are linked to decreased activity in the right ventral prefrontal cortex, while depressive episodes are linked to decreased activity in the left ventral prefrontal cortex.

Doped Up on Dopamine

People with bipolar disorder have differences in their brain chemical balance, too. When a person is manic, there's an increase in dopamine signaling, and artificially enhancing dopamine signaling can lead to a mania-like state. Some people think that cyclical changes in dopamine signaling might be part of why bipolar disorder creates such huge emotional extremes—increased signaling during mania triggers a reduction in the brain's sensitivity to dopamine, which goes too far in the other direction and pulls the person into depression.

NEUROTRANSMISSIONS FILM CORNER:
DEPRESSION

Hollywood holds a lot of influence when it comes to current cultural beliefs surrounding mental illness, including our understanding of depression. Here's how a few of Hollywood's greatest hits fare in their accuracy.

It's a Wonderful Life (1946)

George Bailey becomes so overwhelmed with problems that he's considering suicide on Christmas Eve. But his guardian angel, Clarence, saves George and shows him his true worth.

Depiction of Depression: ★★★★☆
George's descent into depression after experiencing high stress seems pretty accurate. Even strong people can succumb to mental health issues.

Plot: ★★★★☆ This movie tanked in theaters, but has become a Christmas staple thanks to daytime TV replays. It's karmic and satisfying.

Little Miss Sunshine (2006)

A dysfunctional family barely holds together on a road trip to California in order to support young Olive's aspiration to win a beauty pageant.

Depiction of Depression: ★★★★★
Steve Carrell as Frank gives an extremely convincing portrayal of severe depression. His recently bandaged arms show us that depression's symptoms can fluctuate, though Frank is clearly not okay.

Plot: ★★★★★ It's a warm and refreshing story about family without being cheap or clichéd. It balances on a knife's edge between heartache and hilarity—and often does both!

Garden State (2004)

After returning home to attend his mother's funeral, stoic Andrew Largeman decides to stop taking his psychiatric medications and meets a manic pixie dream girl who gets him out of his shell.

Depiction of Depression: ★★★☆☆
It portrays the numbness of depression and the importance of forgiving yourself, but demonizes prescribed antidepressants as a shameful and harmful thing.

Plot: ★★☆☆☆ The film's unconventional style really stood out when it first came out, but it hasn't aged well. *Garden State* will likely slip into that vault of nostalgic, but flawed movies.

Silver Linings Playbook (2012)

Pat Solatano tries to rebuild his life after being released from a mental institution. As he struggles to reunite with his estranged spouse, he meets depressed Tiffany, and they learn to help each other cope.

Depiction of Depression: ★★☆☆☆ Anyone would be depressed after losing a spouse, but Tiffany's condition seems more like emotional dysregulation disorder than major depressive disorder.

Plot: ★★★☆☆ This movie feels like an updated version of *Garden State* and I wonder whether it will age in the same way. In my opinion, it was fairly predictable and ultimately forgettable.

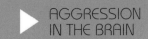
Aggression and fear are pretty primal sensations; being able to fight-or-flight is key to survival for most species. So, where do these evolutionarily ancient emotions come from, and why is it that sometimes you want to punch your brother right in his stupid face?

Hold Me Back, Bro

The word *aggression* means a few different things, so here we're talking specifically about social interactions where there's intent to harm—emotionally or physically—another being. When we think of aggression, we usually think about things like dudebros trying to start fights outside of bars, but it also includes things like chimpanzees fighting for a spot in the dominance hierarchy, badgers defending their territories, or male elk fencing for the affections of a female. Fear, on the other hand, is an emotional response to perceived danger that leads to pretty rapid behavioral changes, like screaming, running, or freezing.

Your Brain on Anger

Like many emotions, aggression is primarily controlled by that lovely limbic system. The hypothalamus plays an especially big role here, as does the periaqueductal gray area; other involved brain regions are the amygdala and the prefrontal cortex. The prefrontal cortex is really important for the impulse control that stops you from actually throwing that punch. Different neurotransmitters also affect feelings of aggression, including low serotonin and alterations in catecholamine systems. Also, testosterone is linked to increased aggression, at least in other species—the connection in humans is less clear.

Don't Be Scared

Fear is similarly controlled by the limbic system, but the amygdala is really where it's at with this one. Studies in animals that have had their amygdala removed have found that they're basically fearless, even walking right up to predators without a blink. If things seem really scary, your amygdala will signal for the release of a bunch of hormones that prep your body for that fight-or-flight response, leaving you feeling shaky and on edge.

WHY DO PEOPLE TROLL?

Trolling describes the act of deliberately upsetting people online and provoking emotional knee-jerk reactions from unsuspecting people. Why do people do this? Some researchers found that self-identified trolls scored higher on dark-triad traits: Machiavellianism, psychopathy, and sadism. In another theory, called the online disinhibition effect, John Suler proposed that six factors contribute to trolling:

1. Dissociative anonymity: "You'll never know who I actually am."
2. Invisibility: "I don't have to say this to your face."
3. Asynchronicity: "I can engage or disengage whenever I want."
4. Solipsistic introjection: "I'm imagining what you're like based on that comment."
5. Dissociative imagination: "I'm just messing with you. It's a game."
6. Minimizing authority: "I'm not gonna get in trouble for it, so who cares?"

With these elements in play, it's no wonder trolling is so prevalent. And if you ever experience it yourself, just remember this lesson: Don't feed the trolls!

other reasonable evolutionary explanations for aggression—perhaps it was to protect one's resources or to defend against attack or to deter mates from cheating. The hard part of many of these theories is that they are difficult to prove because much of the evidence is lost to the sands of time.

Violence Begets Violence

A word of caution when thinking about these evolutionary roots: Just because we can aggress doesn't mean we will. Going berserk in every situation is not evolutionarily advantageous, especially not in the modern era. There are consequences to such actions and, as such, it's important to remember that we humans are also socially aware creatures who have control over our behaviors. Although aggression may be partly determined by genetics, that does not excuse malicious acts that harm others.

Everyone gets angry sometimes, and aggression seems to be just part of human nature. Like it or not, we're a violent bunch. Humans and chimpanzees are the only two species known to coordinate raids on neighboring tribes for the purpose of killing. But if aggression is an intrinsic human quality, it makes you wonder: Why did we evolve to be so aggro?

Survival of the Angriest

Evolutionary psychologists have proposed several theories to explain why humans engage in such horrible acts like war, genocide, and starting a tavern brawl. Quite simply, it all boils down to fitness—if it's part of our DNA, then aggression must promote the survival of our genes. For example, some researchers contend that male aggression, in particular, was used to dominate other males in order to protect a mate, and perhaps to dominate that mate, too. (Yikes!) Female aggression, on the other hand, which often gets described as covert and indirect, may have developed as a way of asserting power over others in order to establish one's status. Of course, we could come up with many

COME AT ME, BRO!

Men perpetrate violent acts approximately ten times more frequently than women. But why? Evidence points to sociocultural underpinnings. In 2018, the American Psychological Association released its *Guidelines for Psychological Practice with Boys and Men* in which researchers wrote, "Traditional masculinity—marked by stoicism, competitiveness, dominance and aggression—is, on the whole, harmful." Now, the APA wasn't demonizing men or male attributes, but pointing out that many men experience pressure to think or act within cultural masculine norms, which stigmatize the expression of certain emotions—like fear and sadness—and elevate other "acceptable" emotions like anger. This limitation leads men to express their anger through aggressive acts that intend to dominate others, which can lead to physical or verbal violence. In fact, men comprise both 90 percent of perpetrators and 78 percent of victims of violent crime. See? Rigid social and cultural enforcement of male gender roles is bad for everyone—including men.

ATTACK OF THE PANIC

Not all stress is bad. Stress is an important physical response that keeps us sharp, and short-term stress can improve our alertness and memory and keep us motivated on a daily basis. But when stress goes from being a motivator to being totally overwhelming, making it hard for you to live your life, that's not normal—that's an anxiety disorder.

Feeling Anxious? Join the Club

Anxiety disorders cover a lot of bases, including everything from social anxiety and generalized anxiety disorder (or GAD), the persistent and unrealistic worry about everyday things, to things like PTSD and panic disorder. These things are super common—fxthe National Institutes of Mental Health estimates that about one in five adults in the US are affected by one. Heck, your very own author Alie is one of them!

Limbic Lovin'

Anxiety disorders, like many other mood disorders, are believed to be a result of changes in brain signaling. Specifically, folks with anxiety disorders tend to show more activity than normal in their limbic system—the complex set of structures deep inside the brain that includes the hippocampus, amygdala, hypothalamus, and thalamus. There's a strong trend for these patients to have a hyperactive amygdala, which might explain a lot.

Personalized Quirks

While anxiety disorders are generally grouped together, they tend to have their own unique flavors, which might be linked to differences in how neuronal signaling is affected by the condition. In panic disorder, amygdala hyperactivity might be caused by less GABA, the main inhibitory neurotransmitter, in some areas of the brain. This could lead to less inhibitory signaling in the emotion circuits, making it harder to control panicky feelings.

Patients with generalized anxiety disorder seem to have larger amygdalas—so their brains have more machinery to process fear information, which then reacts more strongly to negative emotional stimuli. PTSD, on the other hand, might be a result of too much excitatory signaling in the hippocampus and amygdala, leading to intense emotional reactions to triggering stimuli. PTSD might also be partially

WHEN PTSD GETS COMPLICATED

When we think of PTSD, most often we think of it as a reaction to a single event, like a car accident or an assault. But when someone experiences prolonged, repeated trauma over months or years—like ongoing abuse, repeated torture, or living in a war zone for long periods of time—they may develop complex posttraumatic stress disorder (CPTSD). CPTSD has many of the same symptoms as PTSD, plus some. These might include difficulty controlling emotions, dissociation, guilt or shame, and loss of one's religion or faith in the world. These symptoms become more severe if the trauma occurs at a young age, or over a long stretch of time and/or is caused by a parent. CPTSD is still a relatively new condition, and due to its intensity and duration, it can be quite difficult to treat. But as we learn more about it, therapy and medication options will continue to improve.

caused by our logical brain regions being co-opted to process emotional information, giving our brain a harder time controlling those thoughts.

In social anxiety disorder, being exposed to images of faces leads to extra activity in the amygdala—so people who are anxious in social situations are processing social information through a layer of fear, making those environments stressful for them. As in PTSD, this might be a result of excessive excitatory signaling in the limbic system.

PTSD: Not Just for Soldiers Anymore

Awareness of PTSD really got off the ground when soldiers returned home from World Wars I and II with "shell shock," exhibiting extreme fight-or-flight responses and panic. While rates are still very high among soldiers who see combat, PTSD can happen to almost anyone after a traumatic event, and nearly 10 percent of people will experience it at some point in their lives. It's extremely common after assault-based trauma, like sexual assault or child abuse, but can also be experienced after a serious accident or natural disaster. In fact, many folks in healthcare fields are worried about a second "pandemic" after we finally get COVID-19 under control (or heavens forefend, face a new pandemic)—the potential for high rates of PTSD among medical professionals who struggled to save lives when the pandemic was at its worst.

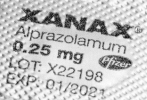

YOUR BRAIN ON . . .

BENZOS

What It Is: Benzodiazepines, or "benzos," are named after the drug's chemical structure, a benzene ring fused to a diazepine ring. Name brands include Valium, Ativan, and Xanax.

What Type of Drug It Is: These psychoactive drugs are used as short-term, fast-acting antianxiety medications.

What It Does: The benzodiazapine category has strong sedating and muscle-relaxing effects, act as anticonvulsants, and can help you sleep.

How It Does This: Benzos bind to the GABAA receptor on neurons and boost the effects of GABA (the brain's main inhibitory neurotransmitter) at the synapse. Increasing GABA signaling dials back overall signaling between neurons in the brain, slowing and calming things down.

What the Risks Are: Benzodiazepines are considered mostly safe for short-term use, but most psychiatrists take a less-is-more approach, and encourage folks to use them only for a couple of weeks or as rarely as possible when dealing with a panic disorder. Because benzos can be sedating, overdosing can lead to deep unconsciousness, and if mixed with other substances like alcohol or opioids, can end up being fatal.

ADDICTION IN THE BRAIN

For as long as humans have existed, so has our love for mind-altering substances. (we're serious. We literally evolved the gene to metabolize alcohol before we even evolved into being humans.) And we seriously *love* to get messed up. So much so that sometimes our brains end up getting hijacked by our drug of choice and we just can't get enough.

Again . . . Again . . . Again . . .

There's no way around it—drugs feel good, no matter how bad they may be for us. Many of the substances that people use activate our reward circuits, dumping lots of neurotransmitters in the brain that make us feel euphoric, in addition to other effects like feeling energized, disinhibited, sedated, and so on. We end up liking the way it feels, so we do it again, and sometimes we keep doing it. When you're using a substance compulsively even though you want to stop, that's an addiction.

I Wish I Could Quit You

Using a substance over and over again can cause your brain to start adapting to the repeated exposures, leading to changes in gene expression and signaling that affect memory, learning, and especially reward. The reward system gets hijacked and a habit forms, hardwiring that behavior into your brain. Now you're stuck in a cycle that can be really hard to break out of. And some substances—like alcohol, nicotine, and opioids—not only hijack your reward systems but also lead to changes in neuronal signaling that make it really uncomfortable or even physically painful for you to be sober, further reinforcing the behavior.

One Step at a Time

Addictions are hard to treat because our brains love rewards, and they also love habits. Therapy and community support can be very effective (more on that in a second), and for some substances, medication can also help. In fact, benzodiazepines, while addictive in their own right, are also a front-line treatment for alcohol addiction, and drugs that mimic the effects of nicotine can be used when someone's trying to quit smoking. Some medications can even block the rewarding effects of the drug, making the substance boring or even unpleasant to use, which can help break the habit.

YOU'RE NOT A BAD PERSON

Pretty much as long as humans have had psychoactive substances, they've had issues with abusing them. Even millennia ago, there were physicians who understood addiction for what it is: a disease. Some people view addictions as a personal moral failing, or a sinful behavior, but often, people really want to stop using the substances—they just can't. It can be hard to see addiction as a disease when it sort of begins with a personal choice (to use a substance), but the reality is that different people have different levels of susceptibility to the condition, and what might be just a couple of casual drinks for one person could be the trigger for alcohol dependence for someone else. Understanding that an addiction is a result of biology and not a result of being a bad person is one step toward destigmatizing and addressing the underlying issues that lead to substance abuse in the first place.

ADDICTION IN THE MIND

While the neuroscientific approach to addiction traditionally views addiction as brain changes that result from chronic consumption of drugs, psychology takes a more emotional, social, and environmental approach.

I Was Predisposed

Most recreational drugs have the capacity for forming an unhealthy habit, but there are multiple factors that play a huge role in the development of an addiction other than its feel-good properties. From an emotional perspective, a person's stress level or history of trauma may lead them to use substances as a way to self-medicate. On a social level, if someone's family or friends abuses drugs, then recurrent exposure to the drug might make a person more likely to become dependent. And environmentally, if the drugs are easy to access and the price is right, this could lead to more frequent use, resulting in those brain changes we were talking about earlier.

I Can Stop Whenever I Want . . .

Of course, addiction is a behavioral issue, too. People keep doing drugs despite the physical or psychological harm they cause, even if the harm is exacerbated by repeated use. And the more you use, the more your body adapts and develops tolerance. This can lead to some of the characteristic symptoms of addiction, like taking more of the substance to get the same effect, using it even when it causes problems in relationships or other life areas, experiencing cravings, and wanting to stop, but not being able.

They Tried to Make Me Go to Rehab

They say that the first step toward recovery is acknowledging you have a problem. After this important phase, most folks recovering from addictions pursue a variety of treatment options, like rehabilitation programs, self-help groups, and individual counseling. Usually, these are used in tandem for ultimate recovery power to address the psychological, social, and environmental factors that perpetuate the addiction that folks want to overcome.

RAT PARK!

Back in the day, scientists studying addiction would place an animal in a Skinner Box (see page 92) and give them unlimited access to drugs. Researchers found that animals would compulsively consume the drug, often to the point of overdosing. The takeaway from these studies was, "If you consume drugs, you'll get addicted and die!" But in the 1970s, psychologist Bruce Alexander studied substance dependence from a different angle. He created an enormous rat utopia that measured over a hundred square feet, with toys, exercise equipment, and water bottles filled with sweetened morphine. He called it Rat Park, and communities of rats were given free reign to do drugs, have sex, and explore at their leisure. Surprisingly, despite having unfettered access to the morphine, rats in this happy, communal environment were far less likely to become addicted to the morphine than isolated rats. Rat Park revealed how one's environment can drastically impact their addiction.

PART
THREE

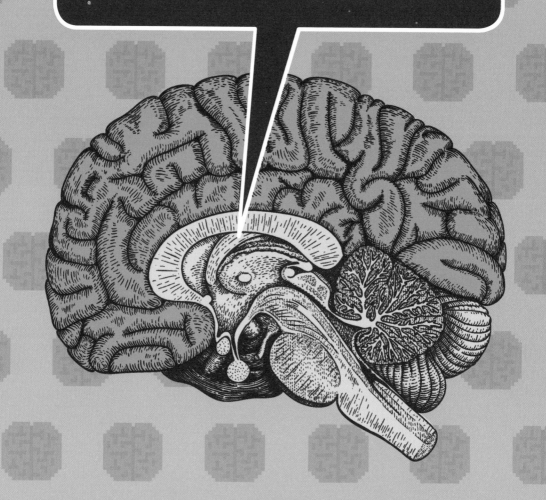

FOREBRAIN

The forebrain is the largest part of your brain, resting atop the midbrain and brainstem and including most of the brain structures responsible for higher-order behaviors and cognition, like the cortex. The forebrain is what processes all our senses, regulates our emotions, and lets us solve problems and make decisions.

In our book, the section about the forebrain is the shortest, but it might be the one with the longest reach. Here, we'll look ahead to the future of neuroscience and psychology, with all the bright possible futures and dark potential dystopias.

We are finally starting to understand how all the complexities of the brain—from the smallest molecules floating between our synapses to the sweeping patterns of blood flow throughout our cortices—come together to drive a living, breathing human. And along with that understanding come new possibilities for creating technology to save lives, better humanity, and expand our futures.

But none of these are quick and easy fixes, and some of them could have pretty negative consequences if scientists and physicians aren't careful. Some of them have already raised hard-to-answer questions about our minds, like who gets to decide what diseases need to be "cured," and whether or should be implanting brain-boosting electrodes in people's heads. Good or bad, the future is coming at us fast; let's take a look at what's coming up on the horizon.

Chapter 16

THE FUTURE OF CLINICAL TREATMENT

The future . . . is now. No, seriously—all this stuff we're gonna talk about is starting to happen *right now*. And it's all pretty amazing.

You've probably figured out by now, two-thirds of the way through this book, that the brain is one big, weird, squishy mess, and we really don't understand it very well. But as we start to figure out how all of it works, one synapse at a time, we're also starting to better understand how we can best take care of it. This chapter is all about the new and innovative ways scientists, physicians, and therapists are helping patients protect their brains and their mental health.

Some approaches in the past were straight-up terrible, like lobotomies. Some were based mostly on guesswork, like SSRIs and early forms of psychotherapy. More recently, we've been able to better tease apart the actual biology of the brain to come up with new treatments and medications and develop more sophisticated, evidence-based approaches for therapy.

This has led to lots of neat breakthroughs, like actually testing and implementing therapies that account for our new understandings of different mental health conditions, and using once-scary psychoactive drugs to treat serious mood disorders. We're also better understanding how our bodies and brains, and all the different microbes and molecules therein, interact with one another to maintain our health—and what we can do to fix things when things are going wrong. Who knows what we'll figure out next?

NOT YOUR GRANDPA'S THERAPY

For decades, "therapy" has looked pretty much the same. Sure, we don't lie down on the couch anymore, like in the days of Freud, but it has otherwise persisted as an in-person, stand-alone service. Psychotherapists have a reputation for being slow to change, clinging to archaic tools and methodologies. However, in the past few years, numerous new techniques and approaches have taken therapy by storm, changing the face of therapy.

We Have the Technology

Unsurprisingly, the internet has completely transformed the mental health landscape. A bunch of tech companies have begun to offer online counseling, text chat therapy, and on-demand services. This has opened up service to a lot of folks who otherwise would not access mental health resources, by making it convenient, lowering costs, and offering 24-hour service. Likewise, your mobile phone offers many apps that track your sleep or diet or mood or health. Using these tools therapeutically can provide more reliable insight into a client's lifestyle than a self-report on the day of the session.

Talk the Talk, Walk the Walk

Believe it or not, more therapists are starting to incorporate the outdoors and exercise into sessions. Walk-and-talk therapy, for example, continues to grow in popularity, and proponents say it enriches counseling sessions because clients seem less anxious, more creative, and less "stuck" in their issues when they're out in the real world. This should come as no surprise—exercise and nature both have well-studied, positive effects on mental health! So, moving in real life might reflect our ability to move forward mentally, too. Don't be surprised if your next therapist wants to take a stroll around the block, do some yoga, or go for a jog.

What's Up, Doc?

Healthcare in many places, particularly the United States, is . . . disjointed. Physical health care and mental health care have usually operated independent of one another, with the exception of a few social workers scattered in hospitals. However, there is a renewed push to integrate behavioral health into primary-care practices—essentially, having a mental health professional in the same office as your doctor. This seems intuitive, given that almost a third of all mental-health-related office visits are to primary-care physicians, and half of patients given referrals to therapists don't make a first appointment. But with our powers combined . . .

A New "Play" Therapy

Play therapy has traditionally been used with small children to help them explore and communicate their inner world. Then, as you grow up, therapy becomes a serious business where you sit and talk out your feelings. But why? Therapy is an opportunity to express yourself deeply and authentically, and, it turns out, play therapy can be a wonderful way for adults to break out of their shell. Modern adult play therapy taps into your creative outlets, like arts and crafts, singing, dancing, or storytelling. But it could also include cooking, photography, or creative writing. In fact, some therapists (Micah included) have used Dungeons & Dragons as a form of play therapy to help people with social anxiety become more comfortable with themselves. Doesn't that sound fun? Roll for initiative!

SHOULD POLICE RESPOND TO MENTAL HEALTH CRISES?

Mental health situations are responsible for almost one in ten calls to police. This means that police officers are often the first to respond to someone in a crisis. Unfortunately, this can have tragic consequences. While only three percent of American adults suffer from a severe mental illness, they make up 25 to 50 percent of fatal law enforcement encounters. This issue is only magnified for individuals with other marginalized identities, who are more likely to be experience violence at the hands of law enforcement officials.

What's Being Done?

Many police agencies have offered psychiatric-response training to interested officers and even hired mental health professionals. However, most agencies struggle to implement the training, and in real-life situations, officers often fail to notice that mental illness is involved.

Additionally, because police feel they have few other options, people in crisis are frequently needlessly arrested and taken to jail. This has led many to question whether police should be involved in responding to mental health crises that they are not well-equipped to handle.

So, What's the Alternative?

The alternative is to create something new. Medical emergencies, such as heart attacks, strokes, and other nonvehicular accidents, are handled by paramedics, not a police officer. As such, one proposal is to create a mobile crisis-response team made up of mental health professionals, community health workers, and peers who have the knowledge and skills to appropriately de-escalate crisis situations and connect folks with resources. No police. With this empathetic approach, we could avoid unnecessary hospitalization, decrease arrests, and, most important, save lives. Could this be the future of mental health crisis response?

A THERAPIST IN YOUR POCKET

The COVID-19 pandemic transformed what many industries looked like—and therapy is no exception. A large swath of mental health professionals who used to conduct sessions only face-to-face have now had a taste of virtual therapy and . . . don't hate it! With younger, technologically literate generations seeking care, teletherapy (which broadly encompasses all virtual forms of therapy, like video, text, or phone) seems like the future. It's convenient, accessible, and just as effective. However, despite these advantages, don't expect therapy to go completely virtual. There will always be a place for face-to-face treatment. Inpatient programs, like those for folks struggling with serious mental illnesses or substance-use disorders, will continue to be necessary and important. Likewise, it's really hard to do group sessions virtually, because you lose that physical presence that is so important to group dynamics. But even so, teletherapy will likely become the new mode of operation for individual counseling.

DESTIGMATIZING MENTAL ILLNESS

Admitting you have a mental health problem and need help can be difficult. But the stigma surrounding mental health can make it even harder. Harmful stereotypes and false beliefs, like thinking people who are mentally ill are "weak" or "dangerous" or "just need to try harder," prevent people from pursuing treatment and lead to feelings of shame. Unfortunately, these negative attitudes are common and have existed for a long time, particularly in Western culture (see Medieval Europe on page 35). They come from the misguided view that individuals with mental illnesses are "different" from everyone else. But of course, this isn't true either. Luckily, this is changing.

An Apple a Day

While older generations whisper about friends who "talk to a shrink" or who "use meds as a crutch," younger generations born after the 1980s have generally embraced mental self-care with open arms and speak freely and honestly about their psychiatric experiences and involvement in therapy. This shift has led advocates to request that mental health be treated at the same level as physical health, both by eliminating stigma and by increasing insurance coverage to get people the services they need. But there's still work to do in this area.

Walk in, See This, Wat Do?

If you'd like to destigmatize mental health issues, there are a few things you can do to help. First, talk about it. Educate folks around you about the realities of mental illness, how common it is, and your own experience. People talking about firsthand experiences of treatment for mental health is the single most effective way to get other people to seek help themselves. Second, check out organizations like the National Alliance on Mental Illness (NAMI) to connect with others and to find resources to support people in your own life.

ISN'T THERAPY FOR "CRAZY" PEOPLE?

Many individuals feel hesitant to go to therapy in the first place because they think that means admitting that something is "wrong" with you. Not so! Therapy is not just for people with mental illnesses. Think of it like going to the doctor. Many folks visit their doc, even when they aren't sick—they might be getting a checkup or a test or need some advice. Similarly, therapy can provide these services for your mental health by offering a nonjudgmental space to talk through problems and receive support so you can live a healthier, happier life. Therapy is a tool to explore yourself and become more self-aware, not a thing that's meant to "fix" you. So, try it out! You might enjoy it a lot more than you expect.

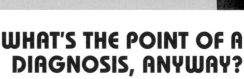

What It Is: Adderall (also called Addys, uppers, pep pills, and study buddies) and Ritalin (methylphenidate, also called Skittles, kiddie cocaine, and Vitamin R).

What Type of Drug It Is: Amphetamine (Adderall) or Piperidine (Ritalin).

What It Does: Adderall is pretty similar in its effects to meth, and can similarly enhance cognition and stimulate the nervous system. Ritalin reduces distractibility and impulsivity and increases attention spans. Both are first-line treatments for ADHD and narcolepsy.

How It Does This: Adderall, like methamphetamine, stimulates the activity of dopamine and norepinephrine in the brain, while Ritalin acts as a NDRI—a norepinephrine-dopamine reuptake inhibitor. Both play roles in alertness and attention, while also affecting other neurotransmitter pathways that help "perk up" the nervous system.

What the Risks Are: Overdoses can result in symptoms and health issues similar to that of other stimulants, including methamphetamine. Adderall and Ritalin are effective for treating ADHD, and are also popular recreationally and as a study aid because they can induce feelings of euphoria and boost energy and focus. But folks with ADHD don't generally feel high when they take their medication; they just feel ordinary, and actually able to focus on what they're doing.

WHAT'S THE POINT OF A DIAGNOSIS, ANYWAY?

When you walk into a doctor's appointment, you tell your physician what's ailing you—sore throat, pain in your side, headaches, feeling tired all the time . . . the doctor takes these data points and fills in the gaps by asking other questions or running tests. Once there's enough information, the doctor considers the constellation of symptoms and gives you a conclusive diagnosis: "Ah, yes—it's dehydration!" Mental health diagnoses work much the same way as physical health diagnoses.

Why Are They Good?
Mental health diagnoses are essentially labels for a group of symptoms that describe what you're going through. This can be extremely useful because, of course, once a professional has correctly identified the issue, understanding and treating that issue becomes a whole lot easier. That doesn't mean that you're put into a box; rather, it provides context to better inform your individualized treatment. Additionally, having a name for what's going on can make

communication with the client, their involved parties, other healthcare providers, and insurance companies much easier and more efficient. Having a diagnosis can also provide comfort by helping folks make sense of their sometimes scary and confusing symptoms.

What's the Problem?
Despite the advantages, diagnoses can also have downsides. While they provide comfort to some, they can negatively impact others if clients have existing stigmas about particular mental health issues or if they are discriminated against by others. Alternatively, if a person holds too tightly to a label and uses it as an identity to excuse or perpetuate unhealthy behaviors, that can be harmful or even dangerous. Mental health professionals need to be wary, too. Certain disorders (like ADHD and borderline personality disorder) can be overdiagnosed, while others (like substance-use disorders and PTSD) go undetected. Regardless, the benefits of diagnosing far outweigh the costs.

STREET DRUGS IN THE DOCTOR'S OFFICE?

While your hippie aunt may have been preaching the values of tripping on 'shrooms for mental wellness since the '70s, thanks to a lot of misinformation, racism, and anti-counterculture sentiment in various world governments, most of the drugs we'd consider psychoactive have been illegal and nearly impossible to study in a clinical setting for decades. Some of that is changing now, which is great news, because some of these mind-altering substances seem to be pretty helpful for treating a lot of neurological and mood disorders. There's even a whole organization dedicated to this cause! The Multidisciplinary Association for Psychedelic Studies (MAPS) is dedicated to supporting research and public understanding of why tripping might be good for you.

The Kush Will Cure Ya

Cannabis, generally better known as marijuana, has quite the sordid reputation (thanks a lot, *Reefer Madness*) but there's a reason this good, good herb has been legalized for medical use in 35 states. The two main components of cannabis are delta-9-tetrahydrocannabinol, or THC, and cannabidiol, or CBD. Both affect the endocannabinoid system, which is pretty multifaceted: It plays a role in cognition, pain perception, and motor movements. Cannabis has been investigated for its potential to treat nausea, improve appetite, and treat chronic pain and anxiety, but the government's resistance to allowing research has made it hard to determine how well it works, at least so far. Probably the most well-studied medical use for cannabis is the utility of CBD for preventing seizures in some severe forms of epilepsy.

Go Ask Alice . . . to Be Your Therapist

The more traditional psychedelics, like LSD and psilocybin (aka magic mushrooms), were touted early on as potential "miracle drugs" because of their profound effects on the mind, including dramatic shifts in perception, mood, cognition, behavior, and even spiritual experience. Beyond recreational use, researchers are working to understand whether these drugs might be helpful for treating serious mental health conditions. Most of the trials so far have been pretty small, but have had promising results—psilocybin has been found to be extremely effective for breaking nicotine addictions, while a single dose of LSD was effective at reducing alcohol consumption in alcoholics. So far, it looks like these drugs (along with psychotherapy) may be helpful in treatment-resistant depression, anxiety, and drug dependence.

The Love Drug

MDMA, also known as Ecstasy, earned that nickname for a reason: It induces feelings of euphoria, energy, and enhanced empathy. These effects are why it's being explored as a means to treat certain mood disorders, and it has even received special approval as a "breakthrough therapy" from the FDA for treating PTSD. Thanks to the ways in which MDMA enhances trust and reduces feelings of fear, MDMA-assisted therapy is being used to help patients struggling with psychological trauma and support the terminally ill as they come to terms with their death.

Special K for Special Kids

Historically, ketamine has been used as an anesthetic, providing pain relief and sedation in medical settings. Recently, it's started to gain a reputation for the dramatic benefits it seems to have in cases of severe depression and suicidal ideation. Clinical trials have found that a single dose of ketamine can rapidly lift depression, taking just a few hours to do what most antidepressants can't do in weeks or even months. Most of the evidence indicates that the effects can also last for weeks to months!

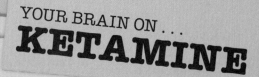

What It Is: Ketamine, aka Special K, Super K, Vitamin K, Cat Valium, Kit Kat, and Purple.

What Type of Drug It Is: Pain reliever and sedative.

What It Does: Ketamine causes a sense of dissociation—detachment from the environment and oneself. It can induce a trancelike or dreamlike state.

How It Does This: Ketamine is thought to block the NMDA receptors, one of the main glutamatergic (excitatory) receptors in the brain, as well as interacting with lots of other neurotransmitter systems, including opioid, dopamine, serotonin, and acetylcholine. Blocking NMDA reduces excitatory signaling in the brain, and is probably what underlies ketamine's dissociative effects.

What the Risks Are: Ketamine is used recreationally for its euphoric, dissociative, and hallucinogenic effects, and it may have enormous potential for treating severe depression, but keep in mind this substance originated as an anesthetic and analgesic. At high doses, users can find themselves falling into a "k-hole"—a point at which a person feels so dissociated that they're unable to interact with the world around them, including other people.

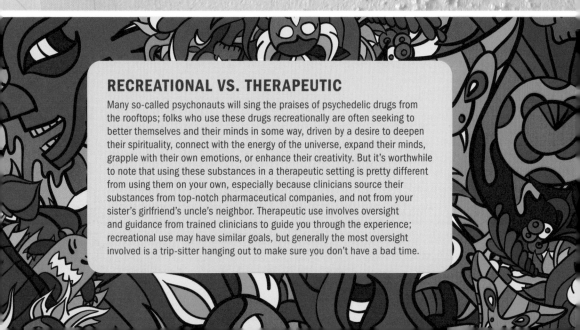

RECREATIONAL VS. THERAPEUTIC

Many so-called psychonauts will sing the praises of psychedelic drugs from the rooftops; folks who use these drugs recreationally are often seeking to better themselves and their minds in some way, driven by a desire to deepen their spirituality, connect with the energy of the universe, expand their minds, grapple with their own emotions, or enhance their creativity. But it's worthwhile to note that using these substances in a therapeutic setting is pretty different from using them on your own, especially because clinicians source their substances from top-notch pharmaceutical companies, and not from your sister's girlfriend's uncle's neighbor. Therapeutic use involves oversight and guidance from trained clinicians to guide you through the experience; recreational use may have similar goals, but generally the most oversight involved is a trip-sitter hanging out to make sure you don't have a bad time.

BACTERIAL BRAINS!

The trillions of microbes in your gut form a whole complicated community called your microbiome. It's kind of obvious that the microbes living in your guts could play a role in your dietary needs and gastrointestinal health, but what might not be so obvious is that the microbiome in your gut can also affect your brain.

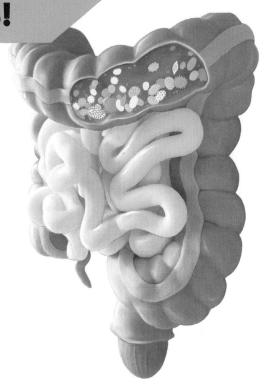

Halt! Who Goes There?

Scientists used to assume that the microbes in your gut couldn't affect your brain, because of the blood-brain barrier. Your brain's blood vessel cells are packed tightly together in a way that keeps your brain's immune system basically separate from the rest of your body to protect against dangerous infections. It's hard for almost anything, including microbes, to get through the barrier, except by serious injury or illness. For a long time, neuroscientists figured that meant they could pretty much ignore the microbiome.

Ew, Emotional Cooties

In 2004, some scientists in Japan looked at how "germ-free" mice—mice that had almost no exposure to environmental microbes—reacted to stress compared to mice that had been exposed to a known set of microbes. The germ-free mice got way more stressed out, and their brains had lower levels of brain-derived neurotrophic factor, a protein important for learning and memory.

The Gut-Brain Axis

It turns out that the relationship between your gut and your brain is a lot stronger than you might have thought. For example, 80 percent of the serotonin produced by the body is made in the gut, not the brain. So, it kinda makes sense that your brain might be affected by what's happening down there. Additional research has found that swapping microbes in the gut can affect behavior (at least in mice) and mood and cognition (in humans). Which leads us to a bigger question: Are your gut microbes *affecting* your brain . . . or are they *controlling* it?

POOP: IT'LL CURE WHAT AILS YA!

You may have heard of fecal transplants (which are exactly what they sound like) as a "miracle cure" for hard-to-treat digestive problems, like obesity and *Clostridium difficile* infections. But what about for brain problems? As researchers have sought to better understand how the gut-brain axis works, they've started to test whether swapping out the microbiome by replacing it with someone else's can help treat psychiatric conditions like depression and anxiety. And the research so far is . . . surprisingly optimistic! In small trials in human participants, studies have found that transplanting microbes from healthy subjects into those with anxiety or depression leads to an improvement in the symptoms of the condition, at least for a few months—but it'll take a while before we know exactly how it works. And . . . probably don't try this at home. You don't know where those microbes have been!

GOING INFLAME

Inflammation is our body's response to virtually everything: Get bitten? Inflammation. Get burned? Inflammation! Get meningitis? INFLAMMATION! So maybe it's not a surprise that inflammation has recently become a target for a lot of neuroscientists trying to better understand what's going wrong in the brain when we're experiencing neurological or psychiatric illness. Inflammation is a pretty critical part of the normal immune response. But as is so often the case, too much of a good thing can start to get ugly.

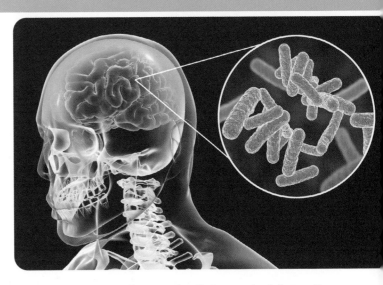

Inflamed in the Membrane
Inflammation in the brain is pretty bad. Called encephalitis (or meningitis, if it's in the membrane surrounding your brain), it's usually caused by viral infection, and it can mess. You. Up. For some folks, there are no symptoms, or just mild ones, similar to the flu. But for others, it can be deadly. That inflammation can cause problems like swelling in the brain, and lead to seizures, paralysis, and even death.

But That's Not All!
In recent years, researchers have started to suspect that mild, less deadly inflammation might play a role in a number of neurological and psychiatric conditions. In some conditions, there are changes in the levels of markers for inflammation, like increased levels of the protein interleukin-4 in bipolar disorder, and increased levels of interleukin-6 in depression and schizophrenia. There's also a growing suspicion among neuroscientists who study

neurodegeneration that excessive inflammation may play a role in conditions like Azheimer's disease, either because the buildup of amyloid beta plaques leads to chronic, damaging inflammation or maybe because inflammation in the body leads to chronic inflammation in the brain and kick-starts cell death.

BUGS IN THE BRAIN
In the last couple of decades, we've started to understand that Alzheimer's disease isn't as simple as "amyloid beta plaques + tau tangles = neurodegeneration." There are a lot of theories on how these well-known biomarkers may be linked to other neurological and physical changes that contribute to this deadly disease. One of them is centered on those itty-bitty bugs in your body. Some scientists suspect that imbalances in the microbiome might cause progressive inflammation in the brain, leading to the disease; in particular, there's some evidence that chronic dental and gum issues caused by pathogenic microbes might contribute to the development of dementia. So, remember to floss!

Chapter 17

CUTTING-EDGE NEUROTECHNOLOGY

We can build it. We have the technology. With neurotechnology bursting onto the scene, we now have the ability to understand and manipulate our brains better, faster, and stronger. Step aside, science fiction, this is *real* stuff!

Technology is evolving more rapidly than ever before, and that means exciting new things for the brain. Editing your genes? Check. Mind-controlling laser beams? Check. Doing research on living brains? Check. Reading your mind with magnets? Prosthetic hands as fancy as Luke Skywalker's? Growing new neurons in your brain and in a dish? Check, check, and check, baby!

We're gaining all kinds of insights into how the brain works, thanks to this new tech. It's even being used in modern medicine, letting us do things we could only have dreamed of just a few decades ago. Some of these approaches might let us get right into a person's genome and snip out a harmful mutation, preventing deadly neurodegenerative diseases. Some of them are letting us look right inside the brain while it's active, helping us understand what's going on up there when we're reading a book or thinking about our crush. Some of them are even letting us reprogram our own cells to become brand-new baby brain cells, in the hopes of doing what was once impossible: restoring function after a devastating brain or spinal cord injury.

All these technologies, and more, are propelling us into the future of neurotechnology—but remember that just because we could doesn't mean we should. Scientists have to be conscious of the ways in which these neat tools could be used for bad, as well as for good.

REAL-LIFE *SPLICE*: CRISPR

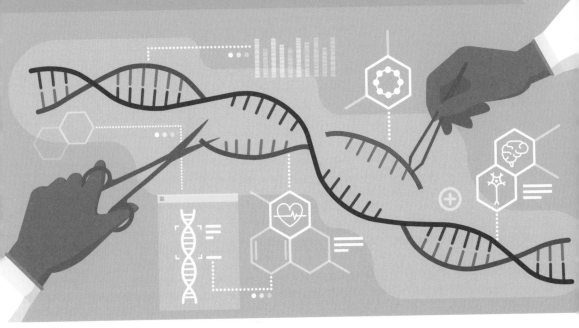

Since we first figured out that DNA is the building block of life, we've made a lot of progress at understanding how changes in the genetic code can lead to devastating diseases. Some conditions, like Huntington's disease, are the result of a single genetic mutation that leads to years of suffering and, eventually, death. It sure would be convenient if we could just . . . hack into the DNA and swap out that bad gene, huh? Enter: CRISPR.

Bacteria Know What's Up

CRISPR is short for "CRISPR-Cas9," the name of a two-part set of defense tools used by bacteria and other microorganisms to fight off invading pathogens (like viruses). The CRISPRs are short, repeating sequences of DNA; think of them like a "Cut here" dotted line within the DNA. Cas9 is a specialized enzyme that acts like scissors to identify and snip through that "Cut here" line. In bacteria, the system chops up the DNA of invaders. In the hands of scientists,

it's a powerful gene editing tool with seemingly unlimited potential.

With Great Power . . .

This tool is a game changer. Before, it took years of genetic manipulation and breeding to create new animal models for various diseases. Now, using CRISPR, it takes just a few months. It's being applied in agriculture to edit plant genomes to make them more drought tolerant or more nutritious, in yeasts to produce biofuels, and in mosquitos to prevent them from transmitting malaria. And perhaps most exciting of all, it's being explored as a way to treat genetic diseases. In a few cases, CRISPR has been used to genetically modify bone marrow cells to activate genes that can overcome the effects of mutations that lead to blood disorders, like sickle cell disease. Scientists have found that, months later, patients who received these modified cells no longer need the standard blood transfusions to treat their conditions.

CRISPR is so powerful that two of the women who first promoted CRISPR-Cas9 as a gene editing tool, Jennifer Doudna and Emmanuelle Charpentier, were awarded the 2020 Nobel Prize in Chemistry.

. . . Comes Great Responsibility

There's a reason why we make horror movies about science run amok. Things can start to get muddy pretty fast if we go screwing around with the wrong things. And it's not just a fear of the horror sci-fi scenario of, like, creating horrifying Franken-pigs with half a dozen human kidneys, mutants that'll eat you as soon as you look at them. It's also a matter of thinking about who's getting access to these treatments, why, and what we're using them for. It might seem obvious that we should use this kind of technology to cure diseases in which a single gene is clearly the cause, like Huntington's disease or sickle cell disease or the BRCA mutation that causes breast cancer. But what about when there's more than one gene causing the condition? What about an edit that gets passed down to a person's offspring? What about a "disease" that might not actually be a disease? Where do we draw the line?

Lines Already Crossed

Some of this is complicated by the fact that we've already crossed some pretty big lines. In 2018, Chinese biophysicist He Jiankui used CRISPR to edit human embryos in order to make them resistant to HIV infection. He was blasted in the international scientific community for shattering one of the critical moral boundaries that existed in genome editing: directly editing the human germline. Now, bioethicists and researchers are scrambling to develop ethical guidelines to inform regulatory policy and ongoing scientific research. The thing that makes it kind of scary is that the tool is almost too powerful. If researchers don't carefully consider the ethics of their actions, it becomes very easy for scientists to throw CRISPR at any problem, without stopping to consider what it means on a larger social or ethical scale. CRISPR holds enormous potential for treating or even curing deadly diseases, but we need to be careful how far we run with it.

NEURODIVERSITY

One of the issues that bioethicists worry about with CRISPR is, who gets to decide what a disease is? Like, sure, it seems obvious we should cure Huntington's disease, a devastating and fatal genetic condition. But what about . . . scoliosis? Or acne? Or ADHD? Or autism? Plenty of autistic folks are totally happy just the way they are, but there are other people outside of the community who see it as a disease. So, who decides when something is "bad enough" to require a "cure"? Think about the ways in which stereotypes about different groups have been used against them to prove that they're "sick"—like women and their hysteria, or the hypersexualization of African American men. There will need to be thoughtful and balanced global conversations on the use of CRISPR in human health to ensure that we don't further marginalize people within an already unequal society.

FRICKIN' LASER BEAMS IN THEIR HEADS

The future is now. We've got the entire World Wide Web in our pockets, chefs have created veggie burgers that bleed like real meat, and scientists can control minds. Wait . . . what? No, that's not a joke!

Awesome Algae

In 2003, scientists at the University of California at San Francisco discovered a protein called channelrhodopsin: a light-sensitive ion channel produced by algae, which opens in response to light, allowing algae to use it to sense the

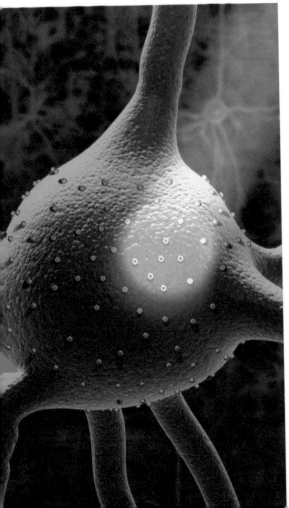

direction of a light source—such as the sun. Scientists immediately realized that this new protein could be the answer to a lot of problems.

Into the Brain You Go

Since channelrhodopsin can be activated by shining a light on it, when it's expressed in neurons, shining a light on those neurons produces an electrical signal. Scientists can create mice that express channelrhodopsin in their neurons, then stick tiny LED lights (or optic fibers) into the brain to focus a light beam on specific structures or cell types. Turning on the light turns on the neurons containing the protein. Researchers can now examine how turning cells on and off affects a lot of different things—including controlling the behavior of an awake animal. Because optogenetics requires only a single gene to create the protein, scientists can target specific types of neurons to make them express channelrhodopsin in order to pick apart the roles of different cells in the brain. New varieties of light-sensitive proteins provide even more control—like changing the speed of the neuron's signal—and some of them can actually suppress action potentials, letting scientists turn off particular neurons within a circuit.

The Possibilities Are Endless

Using optogenetics, neuroscientists have been able to discover some pretty fascinating new stuff. Some researchers have identified a circuit in the amygdala that's involved in how we learn to fear something, and they've also been able to tease apart some of the circuits that are affected in movement disorders like Parkinson's disease. Some scientists are even trying to use optogenetics to help human patients by developing new kinds of pacemakers that use light to regulate the heartbeat. That's the beauty of a technology like optogenetics—it can be used for so many different things.

BIG DATA, BABY

The brain is a biological entity, but it's frequently likened to a machine, or a computer. With billions of neurons and trillions of synapses, one could argue that it's essentially just a crazy-powerful CPU hanging out in our heads. Maybe it's no surprise that as our ability to build external computers has grown, so, too, have the ways we use math to model the brain.

Theoretically, It Could Work . . .

Computational neuroscience is sometimes called theoretical neuroscience, because it's generally conducted with theoretical models and mathematics, as opposed to digging into real, living brains. Researchers take numbers based on real-world phenomena, like the physics that underlie how neurons send signals, and crunch the numbers to model how a neuron might react if given different kinds of inputs or different numbers of outputs. They can also model how neurons grow, how they form connections with one another, and even how whole networks of neurons work together to drive behavior.

But Why, Tho?

Just like any other approach to neuroscience, using what we already know about the brain to build models of its component parts—or even the whole system—is just another way to uncover all the stuff we haven't figured out yet. Using a computational approach can complement work coming out of what we call a wet lab, which is the stereotypical kind of lab you probably think of, with beakers and microscopes and stuff. Thanks to the power of computers, researchers provide new evidence and information to inform next steps in other kinds of research. But don't worry; we're not ready to stick you in the Matrix yet.

DREDGING THE BOTTOM OF THE DATA

When you deal with vast amounts of information, it's not hard to find some kind of result—even if it's not the one you were originally looking for. This has been a huge issue in psychology and other fields when researchers, who take seriously the publish-or-perish mantra, analyze their data from hundreds of angles to find any pattern that is statistically significant, even though it may be totally wrong. This practice is called "p-hacking," where the p stands for the level of significance and represents the probability that your result popped up by chance. Normally, you want your study to have less than a 5 percent chance of being a fluke ($p < 0.05$). But if you have tons of data and if you're willing to test enough hypotheses, you're pretty much guaranteed to find something significant. So, don't trust every big study you read! They may be fooling you with statistical witchcraft.

RESEARCH IN LIVING HUMAN BRAINS

Did you know you can donate your brain to science . . . while you're still alive? It's true! Thanks to the miracles of modern medicine, researchers are finally able to collect and culture pieces of real, actual, living human brains. It's *aliiiiiive*!!!

Braaaaaiiinsss

It's been pretty challenging for scientists to get a clear picture of how human brains function, because so far, they've relied mostly on animal models—and, well, there are a lot of differences between mice and men. Even if you want to use human brains, the tissue is super fragile, so it starts to degrade pretty much the minute you die, making it much harder to donate than a heart or a lung. So, where do you get living brains?!

Who's Donating?

Typically, living brain donors are patients who are undergoing surgery for epilepsy or to remove a brain tumor, so these procedures aren't exactly simple. Often, doctors have to cut out one or two very small (like, the size of a marble) pieces of healthy tissue to get their tools into the right spot. Usually, any tissue removed gets destroyed as medical waste. But in some places, scientists are standing by with their test tubes, eager to swoop in and grab it.

Gotta Go Fast

As soon as the tissue has been removed from the living brain, it will start to break down, so scientists have to rush from the OR to the lab to start processing the tissue. Then, they have a day or two to study the activity and function of the brain cells within their little ball of tissue, which is giving us some really exciting new insights about what makes the human brain unique.

Yes, We're Very, Very Special

We've already learned a few neat things that wouldn't be possible with postmortem (dead) brains—like how human brains express really different patterns of serotonin receptors when compared to mouse brains, which researchers think might explain why some experimental medications for conditions like depression seem to work in mice but not in humans. Researchers will have to be careful, though; once again, these kinds of experiments open up an ethical can of worms. Like how much "say" donors have in the use of their brain tissue, and how much of the brain can you take before it becomes too much like a "real" brain?

3D-PRINTING YOUR BRAIN

Believe it or not, while we don't recommend cracking open your skull and digging around in there, it is possible to touch your own brain. Doctors have recently combined the technology of fMRI and 3D printing to create amazingly accurate replications of patient brains. Why would they do this? Well, some doctors use the 3D-printed brains as practice dummies for complex surgeries. As you can imagine, it's much safer to train on a plastic model than on a living, breathing human. Other doctors and researchers have used 3D printing to better visualize brain anomalies, like lesions or tumors. As imaging and printing continue to improve, they could become vital diagnostic tools. And if you're just a nerd like us and can get your hands on your structural scan, you can 3D-print your own brain!

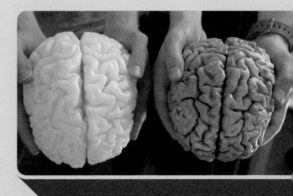

MAGNETS AND THE MIND

Mind reading still sounds like something out of a fantasy story or a sci-fi thriller. But in some ways, we're surprisingly close to being able to read your thoughts, thanks to some big fancy magnets and a lot of electricity.

Functional What Now?

Functional magnetic resonance imaging (what a mouthful) or fMRI machines, let scientists track the blood flow of the brain. Since blood flow correlates with activity, they can figure out which areas of the brain activate when you're doing a math problem, looking at a sad picture, or even smoking salvia. It does this with some complicated physics and a big, doughnut-shaped electromagnet. It can't detect the activity of individual neurons firing, but it can tell where your brain is sending more oxygen, which is a marker for high rates of activity in a given area. Using fMRI, scientists have been able to identify 180 distinct brain regions, including one whose main job appears to be processing faces!

Too Good To Be True

Researchers depend on complex computer programs to analyze fMRI data and figure out if changes in blood flow are significant or not. Back in 2016, a new study made waves, claiming that scientists might be getting it wrong more often than they should, due to a mistake in the analysis that led to abnormally high false positivity rates. A lot of folks were concerned that this meant that 15 years of fMRI data were basically useless; but really, discovering this error is a good thing. With this new awareness, programmers can design better software for the future, and scientists feel encouraged to share more of their data for easier double-checking with newer analysis methods. This is also a good reminder that we can't rely on any one method for telling us how the brain works. We've gotta come at these problems from all angles if we want to be sure that we're really getting everything right.

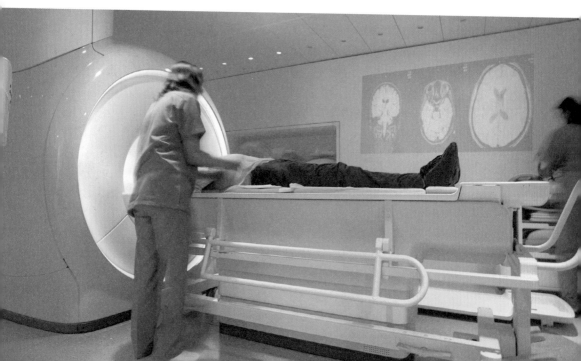

GIMME A HAND

In the Star Wars universe, daddy issues take at least a couple of movies to resolve—but losing a hand is no big deal. Just a few scenes after a major amputation in a dramatic showdown with Daddy Darth Vader, we see Luke Skywalker snapping a panel closed on his brand-new prosthesis, and it's as if he never lost his hand. Like lightsabers and hyperdrive, this advanced medical technology is light-years away for us here on Earth—or is it?

These Are the Prosthetics You're Looking For

Neural prostheses are biomedical devices that use neuroscience and engineering to complete or replace damaged or missing systems—these can include sensory systems, like sound and touch, as well as motor systems, like walking or throwing a ball. There are a wide variety of devices out there for people to use with varying degrees of integration with the nervous system, but when we're talking about Luke Skywalker's hand, we're talking about a true brain-computer interface. That is basically exactly what it sounds like: a direct connection between the brain and a mechanical prosthetic, using the brain's own electrical signals to provide information to control the movement of the device, and ideally sending electrical signals back to the brain to provide sensory feedback.

I've Gotta Hand It to You

We're not quite to Star Wars–level advancement with our neuroprosthetic devices, but we're surprisingly close. Scientists have built computerized prosthetic hands that can be controlled with electrodes implanted into the nerve fibers of the arm, with enough sensory sensitivity that users can grip and manipulate objects without looking at them—even things as delicate as raw eggs! Right now, researchers are working on developing more finely tuned, sensory-feedback systems that will make the device feel more like a "real" hand. By being able to "feel" a prosthetic hand, users can operate their new limb more naturally, and maybe even feel more ownership of the arm.

GROWING NEW NEURONS (IN YOUR HEAD AND IN A DISH)

One of the biggest challenges neurologists face is the fact that once you've killed a brain cell, it's gone. If you suffer from a bad concussion, a stroke, or a spinal cord injury, there's a pretty good chance that that's it—you'll experience some recovery, but what's done is done, and the dead neurons can't be replaced. But with some recent discoveries about the adult brain and some neat new tech, we may soon have better therapies for all kinds of brain injuries.

The Adult Baby Brain Cells

For a long time, there was a pretty firm belief that you're born with all your neurons, and that as you get older, some of them will die, and some of them will adapt, but you'll never grow new ones. Well, that was wrong. The adult brain *does* grow new neurons—at least in the hippocampus, the brain structure responsible for the formation and recall of new memories. We're not exactly sure how the brain integrates these new cells, but we think it's probably important for learning and memory. (Go figure). But the hippocampus doesn't generate enough new brain cells to fix major brain injuries, so what else ya got?

Planting Brain Seeds

New neurons in the hippocampus come from neural stem cells: specialized cells that can divide and mature into a variety of kinds of brain cells. Stem cells for different cell types can be found throughout the body, and these cells represent a major avenue for potential therapies in all kinds of disorders and conditions. We have already successfully transplanted stem cells in damaged eyes to restore vision and in the cochlea to regrow hair cells and restore hearing. Currently, scientists are conducting clinical trials to see if the technology can be extended to repairing serious brain and spinal cord injuries.

Chapter 18

THE CYBERPUNK DYSTOPIAN FUTURE

Where is the future taking us? Will it look more like *Blade Runner*, *The Matrix*, or *Star Trek*? And will we still get to wear cool sunglasses, regardless of which future we end up in?

There are some things about cyberpunk settings that seem pretty neat, like flying cars and powerful virtual reality systems, but other aspects seem less appealing, like the robot apocalypse or a total lack of any natural spaces or the oppressive social order. In some ways, the state of neurotechnology is a little bit cyberpunk: high tech, but with a dark underside.

The unknown can be scary, and it's hard to know how the things we create today might end up getting used in the future. Could the creators of the internet have predicted the challenges of political echo chambers and rampant cries of "Fake news"? Did the person who built the first

computer suspect that we might one day try to upload the human brain? Probably not. Which presents a new question: Should we avoid some of these new inventions if they might cause ethical conundrums in the future?

As is often the case in cyberpunk stories, there's no clear-cut answer, just a bunch of folks doing the best they can with what they have, trying not to incur some horrible unforeseen consequence. But while several of these topics seem pretty dystopian, a lot of it is pretty optimistic, too. So, hang on to your tiny *Matrix* sunglasses, kids; you'll need 'em for the bright future.

FAKE NEWS!

While we might joke about "fake news," it's no laughing matter. Disinformation has become a genuine threat by reducing confidence in real journalism and propping up false information. You can't deny it—misinformation spreads across the internet like wildfire, and might even spread more quickly than real news. But why?

I Don't Believe It

Humans are not very logical beings—but we think we're logical. This causes some issues, particularly when it comes to changing our minds. For example, in one study in the '70s, high school and college students were asked to compare two suicide notes and identify the real one. Most students did fine—sometimes they could identify the real note, but not consistently. After the task, researchers told some students that they were really good at spotting the real note, while other students were told that they were really bad at the task.

Later, the researchers revealed that they lied and clarified that, really, everyone did just fine. But strangely, the students still thought that they were either better or worse at the task than they actually were. The lie stuck with them. So, see? Even when we are given information that should adjust our beliefs, we have difficulty letting go of what we initially felt or understood.

It Just Feels Right

Why is it so hard to change our minds? One contributing factor is cognitive dissonance, which is the idea that people feel uncomfortable when they have inconsistent beliefs or make inconsistent choices. As such, they take actions to resolve this "dissonance," which can result in inaccurate or biased thinking. For example, if you're a chain smoker, you've probably heard that smoking is bad for your health, yet you keep smoking. That clash makes your brain feel bad, so you resolve this dissonance by either not smoking or by ignoring the medical implications. But if you want to keep smoking, then you will likely engage in motivated reasoning, which is a biased way of thinking that helps you reach a conclusion that you want to be true. So you might say to yourself, "Well, I haven't had any health problems yet, so it's probably fine." This can then lead to confirmation bias, which is our tendency to be more convinced by information that supports our beliefs. So, if you see any article or post that says smoking isn't as bad as everyone says, you're more likely to believe it and seek out similar evidence to prop up your view.

Everyone Else Said It Was True

These psychological patterns show that our brains can be pretty easily led astray by misinformation. In the era of social media, it gets even more problematic. There's so much information being flung at us, it's hard to tell which sources are credible and which ones are unscrupulous. Social media companies are motivated to promote whatever gets the most traffic and attention, which isn't necessarily what's true. But the more popular a post becomes, the more likely you are to see it more than once, which makes you more prone to the illusory-truth effect. This is the idea that we tend to believe information we're exposed to repeatedly, whether or not it's true. This effect makes sense—when a whole bunch of people keep talking about the same story, it seems to have more credibility than if you heard it from just one random dude.

They Said WHAT?

But this doesn't completely explain how and why fake news goes viral. These headlines get shared for a reason, and it even seems like fake news gets shared more than real news. A study in 2018 showed that fake news on Twitter spread farther and faster than the truth. We're not exactly sure why this happens, but it appears that fake headlines provoke more intense emotions, like surprise or outrage, which cause people to share them more than real news.

Only You Can Prevent Fake News

Disinformation is not going away anytime soon. So, how can we resist the influence of fake news if it's everywhere and spreads so easily? On an individual level, look more carefully at news sources and think critically about them. Dig deeper into news outlets and authors to understand what biases they may have or how they did their research. Also, when you see a surprising headline, take a second to reflect on it before you decide to share it. Shocking stories might seem important to amplify, but they're not necessarily true. Luckily, several social media companies have started to tag headlines with warnings if third-party fact-checkers have found them to be dubious, which has shown some efficacy. But stay vigilant!

YOUR AI THERAPIST

In 1964, MIT computer scientist Joseph Weizenbaum created ELIZA, a (relatively) simple computer program that simulated conversation. It mimicked the speech patterns of a psychotherapist by parroting things back to the human, making it seem like ELIZA understood more than it actually did. Many people who interacted with ELIZA were convinced it had intelligence. Since then, artificial intelligence has come a long way, thanks to the rise of neural networks and machine learning. At this point, we have AI that can pass the Turing test, which has led people to wonder, could AI chatbots be used for therapy? This is not science fiction anymore. There are now AI-driven therapy apps like Woebot, Youper, and Wysa that use techniques to help people take control of their mental health. Despite the amazing promise these apps hold, at this point they're all . . . limited. While these digital tools can work really well for specific purposes, don't expect current AI therapy to act as a replacement for the real deal. But let's talk again in a decade or two!

For most people, video games have served as a fun diversion. After all, who doesn't like to escape into their imagination to blow off a little steam? But the recent rise of virtual reality has made these experiences more immersive, more interactive, and more "real."

Make It a (Virtual) Reality

The reason it feels so real is because VR mimics our physical surroundings by feeding us sensory information that fools our visual and auditory systems. Our eyes and ears work the same whether we are in the real world or a virtual one. When we simulate an experience in actual reality—like having stereoscopic vision that moves naturally with your head or dynamic sound that seems to come from a particular direction—VR can magically transport us to a different place that feels very realistic. The more closely a virtual environment can simulate the real world, the more immersive it becomes. And by taking advantage of our biology, VR can be used to develop novel, unique experiences that closely resemble reality. As you can imagine, researchers and mental health professionals have taken a look at this technology and said, "Hey, I might have a use for this."

Immerse Yourself

From a research perspective, VR opens new doors. When scientists want to run experiments, they attempt to control as many factors as possible in order to reduce the number of extraneous stimuli that might affect results. The problem with this is that you're then left with a very unnatural environment: an empty white room with a chair and a table. Oh, no—now people will act differently because they feel uncomfortable! But VR could fix that. It can allow researchers to completely control the virtual setting while also providing natural, context-rich scenarios that help participants behave like they're "in the wild," which would yield more accurate results.

Face Your Fear

The immersive nature of virtual reality has demonstrated real therapeutic use for people suffering from anxiety disorders, too. One of the most effective forms of therapy for these disorders is called exposure therapy, which involves "exposing" people to thoughts, memories, or situations that are viewed as frightening or anxiety-provoking. For example, if you were violently attacked in a dark parking lot, a therapist might make you mentally visualize that same situation or might physically bring you to a parking lot in order to work through the fear. But it can be unsafe or difficult to do this kind of real-life exposure during sessions, and clients may not feel comfortable doing it on their own. However, using virtual reality exposure therapy (VRET), therapists can immerse clients in a computer-generated virtual environment programmed to help the person confront their feared situation in the comfort of an office. It's shown significant success for a variety of issues like PTSD, claustrophobia, fear of driving, arachnophobia, social anxiety, and more.

The Future of VR Healthcare

Beyond exposure therapy, researchers have started to explore using virtual reality in the treatment of other mental and physical health issues. Medically, VR has been used to help with rehabilitation after a stroke and with Parkinson's disease patients to improve quality of life and improve daily living skills, and it's even being explored for treating postoperative pain. On the mental health side, virtual reality is being studied for a wide variety of conditions, including helping autistic clients improve their social skills, with depression by letting adolescents literally fight manifestations of their symptoms, and, most interestingly, with gender dysphoria by allowing transgender individuals to embody their gender identity through virtual gender-affirmation and sex-reassignment therapy. This is still in its infancy, but VR could become more than just an expensive fad gaming toy!

NEUROTRANSMISSIONS FILM CORNER:
VIRTUAL REALITY

What if you could fall into a virtual world that was so real that your actions in the digital realm had real-life consequences? This is essentially the premise of all movies about VR, but how realistic are any of them?

The Matrix (1999)

Computer hacker Neo uncovers a forbidden underworld, but he gets more than he bargains for when he discovers the shocking truth—he has been living in a virtual reality his whole life.

VR Realism: ★★★★☆
It'd be *real* hard to make a simulation so realistic that it's indistinguishable from reality. But if you did, you'd probably need a hookup directly to the brain, like in the film.

Plot: ★★★★★ This movie is on my could-watch-anytime list. See it if you haven't yet.

Ready Player One (2018)

While the real world gets ravaged by global warming, a geeky nobody named Wade escapes into a virtual universe called OASIS to claim fame and fortune.

VR Realism: ★★★★★
While the level of immersion seems out of reach right now, VR could get there . . . eventually.

Plot: ★★☆☆☆ Fun for the '80s nostalgia, but the plot is shallow and Wade is a bland stand-in surrounded by more interesting characters.

TRON (1982)

When talented computer engineer Kevin Flynn finds out his work has been stolen, he tries to hack into his company system but gets transported to a digital world.

VR Realism: ★★☆☆☆
We'll never be pixelated and teleported into a video game.

Plot: ★★★☆☆ Despite the story having some weaknesses, this film was way ahead of its time in terms of CGI.

Being John Malkovich (1999)

Craig Schwartz, an unemployed puppeteer in New York, discovers a portal that leads into the mind of actor John Malkovich. But it doesn't take long for things to go wrong.

VR Realism: ★☆☆☆☆ This one was added just for fun, but it's probably pretty unrealistic we'll ever be able to occupy another person's mind anytime soon.

Plot: ★★★★★ It's a quirky, but deep exploration of existential, moral questions that leaves you broken—in a good way.

A GHOST IN THE MACHINE

With all of the incredible technology we've been developing, and the ever-expanding potential of our computer systems, it seems like it's only a matter of time before we all get plugged into an ethernet connection and have our brains uploaded to the Net—or better yet, a robot body. So, what's standing between us and that particular dystopian future?

Billions of Connections

Scientists think there are about 100 billion neurons in your head, capable of forming probably about 100 trillion synapses — that's 100,000,000,000,000 connections in your brain for transmitting all your thoughts, hopes, and dreams. That's . . . a lot. Estimates of the brain's memory storage vary, but we think the capacity is around 2.5 petabytes (about 2,500 terabytes, or 2.5 million gigabytes). That'd take a lot of computer processors, for sure, but it's not insurmountable. The problem is that even if we have a lot of storage capacity lined up, we still don't know exactly *what* needs to be stored.

Connecting the Connectome

Many scientists feel that before we can figure out how to upload our brains to a computer, first, we've gotta map out all those connections. A full map of all neuronal connections in the brain is called a connectome. It's only been in the last decade or so that scientists have been able to create a complete connectome for the humble *Caenorhabditis elegans* worm, which only has 302 neurons and 7,000 synapses. Scale that up to the size of the human brain and, uh . . . this might take a while.

More Than the Sum of Our Parts?

We can count how many brain cells there are and how many connections they make, but we're still not quite sure where, exactly, our consciousness and memories are housed, nor how all the parts come together to make us, us. The current technologies we have to map out the brain's connections mostly rely on snapshot views of the synapses at a particular moment in time, usually by freezing the brain at the moment of death, but we don't know if that actually preserves stored memories. Freezing doesn't capture all the electrical and chemical activity of the brain, particularly the individual molecules jostling around in those brain cells, which might (for all we know) end up being the *real* key to the complex uniqueness of any given individual mind. So, sorry—no computer AI immortality for you.

NEURAL LINKUP

Ugh, Neuralink. Leave it to Elon Musk to spend a bunch of money hiring people much smarter and more talented than him to create a fancy, flashy scientific device for him to name. At least the rocket ships and cars he pays for are useful. But Neuralink . . . well, let's just say we've got a way to go.

You Probably Had the Right Idea

Musk's whole jam with Neuralink is to create a brain-machine interface device that lets humans achieve "symbiosis with artificial intelligence." He believes that AI will eventually exceed human intelligence, and it would be better if we all made it a part of our *own* brains before it takes us all out, Skynet-style. To do this, he's proposed an implantable system he calls "neural lace" that acts as an additional layer on top of the human cortex. By 2020, he already had scientists testing these devices in humans to support his goal of allowing "anyone who wants it to have superhuman cognition."

But the Wrong Niche

The idea of an implantable device isn't crazy; in chapter 17, we talked all about cool new tech that allows users to control prosthetic devices using their brains thanks to implanted electrodes. But these devices are enormously expensive, extremely complex, and only advanced enough to restore some partial degree of function and, if a person is lucky, some sensation. Controlling a machine with your mind is either noninvasive and, therefore,

not super precise, or it is precise and requires an invasive, sometimes dangerous procedure to install. This isn't a technology we should just plug into people's heads for the fun of it.

Maybe Someday

Most neuroscientists agree that while what Musk proposes isn't completely impossible somewhere down the line, it's far beyond the scope of the technology we have available. It will take decades of picking apart the signaling and structure of the brain to understand it all, not just a fancy sewing machine to stick thousands of electrodes into your poor, delicate brain. What Neuralink *is* currently capable of isn't especially innovative, because researchers have been developing these kinds of advanced devices for decades. So, maybe Musk should pour some of his money into the labs of scientists already working on this research, instead of overhyping his tech and shooting Teslas into space.

CAN YOU HEAR ME NOW?

Our better understanding of neurological conditions and more advanced neurotechnology has led to devices that can restore not only missing limbs but also even missing voices. In the past, people who were nonverbal for one reason or another had extremely limited options for communicating with others, and some of them were pretty bad.

I Never Said That

Take facilitated communication, for example. A method popularized for aiding communication in autistic people and others with nonverbal-communication disabilities, the method involves a "facilitator" who "guides" a disabled person's hand to "help" them type out words. This technique has been widely criticized because research has shown that the words being typed are usually coming from the facilitator, and not the nonverbal person, even when the facilitator don't realize it. Kind of like a particularly insulting type of Ouija board. Luckily, we're now developing better alternatives, like speech-generating devices that can be easily tapped by a nonverbal person to produce words.

I'm Still in Here!

Another situation where communication has been extremely difficult is in locked-in syndrome, in which a person appears to be totally unconscious, when in reality they are fully aware. For a long time, communication options for these folks were pretty limited, amounting to having an assistant recite the alphabet letter by letter and the locked-in individual blinking to indicate when the correct letter is reached. Today, we have electronic communication devices that can track people's eye movements and translate them to speech with a computer voice prosthetic, allowing individuals living with the condition to communicate more easily.

BRAIN DEATH

Thanks to lifesaving technology, the line between life and death is kind of murky, particularly in cases where the body might look fine, but the brain definitely isn't. In 2020, as part of the World Brain Death Project, a group of international experts in neurotrauma and critical care put together a new set of guidelines to help other physicians determine when a patient is truly brain dead, meaning the patient has no hope of recovering brain function and should be removed from life support. Their guidelines address different methods for testing brain function, different *kinds* of brain death, and even who should be considered qualified to diagnose brain death. Because this work has enormous legal, ethical, and financial implications, it's not an easy task, but it is an important one. You don't want to mess this up!

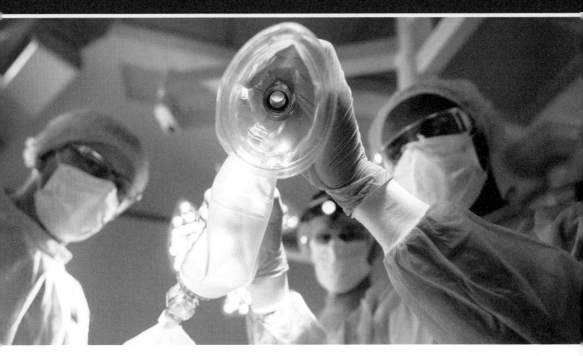

When you need some kind of major surgery, doctors will usually use general anesthesia to keep you unconscious for the duration of the procedure. These drugs typically mean that you have no recollection of the procedure after the fact. But it turns out that some folks—around one person in every 20,000—are awake during their surgeries. This is because the patient hasn't received quite enough of the anesthetic to totally knock them out. It has mostly been a problem when surgeons use paralytics to aid in the surgical procedure, because patients aren't able to signal that they're awake, which is terrifying. Recently, there have been efforts to educate both surgeons and their patients about the possibility of anesthesia awareness to ensure that anyone who experiences it will have access to any additional care they might need to process the experience.

DON'T GO TOWARDS THE LIGHT!

People who have near-death experiences describe feeling sensations like floating, warmth, leaving their body, and meeting spiritual beings. But why is it that lots of people who almost die feel the same things, and what does that say about the brain? It's hard to study so-called NDEs because most people aren't exactly standing next to an MRI machine when they're just about to die, but we have a few theories. Some people think it might be related to a lack of oxygen in the brain as you're dying, or that the brain starts to hallucinate as it dies. Many people think it's due to brain damage as you get close to death; different brain regions start to shut down or malfunction, causing abnormalities in signaling that are perceived as those classic NDEs, like your life flashing before your eyes.

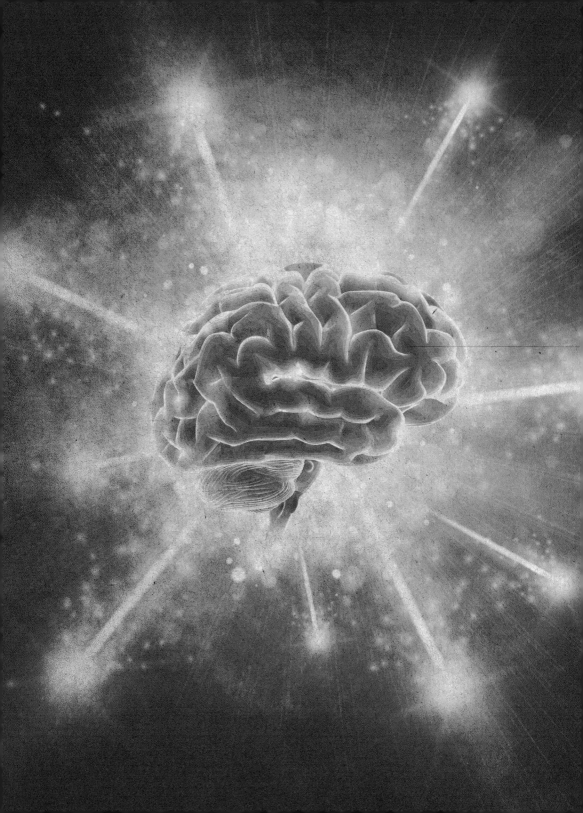

Chapter 19

BOOSTING YOUR BRAIN

Does your brain need supercharging? Lots of folks out there claim to be able to boost your brainpower—but what's legit and what's bunk? What does it actually take?

It's not all doom and gloom in our cyberpunk future, despite the pandemics, social inequality, and climate change. Some of the advances we're making in understanding the brain are leading to opportunities to actually improve and expand our minds beyond what we thought was possible, which might lead us into a better world.

Strangely, even with all the cool neurotech available to us, it sort of seems like some of the most effective tools are the ones we've had all along. Old brain-stimulation techniques, revamped for the 21st century, are now valuable treatment options for people with difficult-to-

treat mood disorders and neurological diseases. Despite the groans of couchbound readers everywhere, physical exercise and a healthy diet are once again turning out to be some of the best options for keeping our brains young, strong, and smart. And a growing awareness of equity and diversity means that we're finally questioning whether the way we define *intelligence* is really useful at all, and looking for better ways to understand each unique person's skills and abilities.

But let's cut through the hype and see what the future has in store for your brain.

A HEALING TOUCH

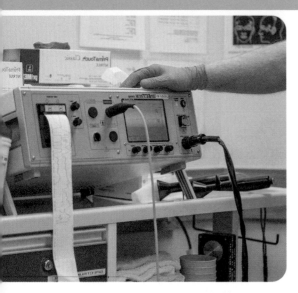

inducing an artificial seizure using electrodes on a patient's scalp, which is believed to lead to changes in brain chemistry that can quickly reverse symptoms of serious mental illnesses, especially treatment-resistant depression. Another more recent approach, known as transcranial magnetic stimulation, or TMS, uses a similar concept, only this time with a magnet instead of electrodes. This allows for more targeted treatment and is thought to have fewer side effects than ECT. But TMS is still new enough that scientists continue trying to figure out the best ways to use the technology.

Sometimes, medications and talk therapy aren't enough to really help patients cope with a debilitating neurological condition, and doctors have to get extra creative to find new treatments. One shocking example of a particularly unusual but somehow effective treatment is electrical brain stimulation.

What a Classic

Probably the most well-known type of electrical brain stimulation is electroconvulsive therapy (ECT). The treatment works by essentially

Really Getting in There

If stimulating the brain from the outside isn't enough for you, how about deep brain stimulation (DBS)? With this approach, doctors actually stick electrodes deep into the brains of patients with conditions like Parkinson's disease and epilepsy, where they can pulse electricity into the relevant brain region to reduce symptoms. DBS is now being explored for psychiatric conditions like depression, chronic pain, and addiction. But since we're still not exactly sure *why* this electrical stimulation is helpful, it might be a while before we figure out exactly *how* the technology should be applied.

WEARABLE TECH

A DBS system is perhaps the most wearable of wearable tech devices, with electrodes and a battery implanted beneath your skin to send electrical pulses into the brain. However, there is a growing industry of less invasive wearable tech to boost your brainpower or improve your mood, consisting of futuristic-looking headbands that send electromagnetic pulses or a constant electrical current through the scalp and into the brain. These pulses are sometimes intended to induce rhythmic patterns of neural activity, which mimic natural brain waves seen in different states like sleep and meditation, or increase blood flow to the targeted brain region to keep things active. While there's not a whole lot of evidence that these devices can really make it easier to focus or help you think, there is evidence that they might be good for things like chronic pain, depression, and even helping people recover after a stroke.

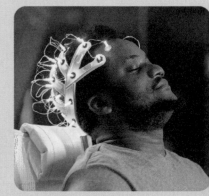

BETTER BRAINS WITH PILLS AND GAMES?

People are desperate to do whatever they can to improve their brainpower and keep their minds younger for longer. Expensive brain games and cognitive-enhancing substances (called nootropics) are touted as one possible solution—but do they work?

Quit Playin' Games with My Brain

So-called brain-training games claim to improve memory, attention, and reasoning skills, and some even purport to prevent the onset of dementia. They supposedly leverage neuroplasticity, the idea that the connections between your neurons can adapt and change over time. Playing brain-training games is supposed to be like exercise for your mind, the way you'd lift weights to exercise your muscles. The problem is that the "brain training" skills you learn aren't really generalizable; playing brain games makes you better at, well, playing only those particular games, not at all things memory related. There is evidence that practicing life skills, like counting change and saying people's names, is good for helping patients with dementia maintain their memories longer, but it's a lot harder to market Practice Reciting Your Grandkid's Names! as an exciting, interesting game.

Super Serums?

Are nootropics—or smart drugs—really the key to unlocking your brain's superpowers? Common examples of nootropics include caffeine, ginko biloba, creatine, nicotine, and L-theanine. Some of these do have measurable effects on cognition, like caffeine, which improves wakefulness, concentration, and motor coordination and may reduce the risk of Alzheimer's disease, and L-theanine, which can help people feel more relaxed. Others, like creatine, might be good for providing extra energy for your muscles, but for your brain? Not so much.

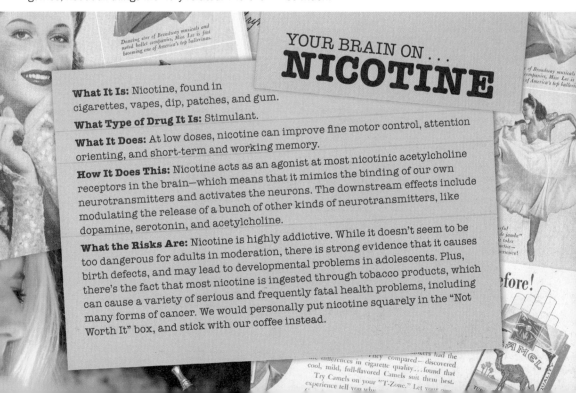

YOUR BRAIN ON... NICOTINE

What It Is: Nicotine, found in cigarettes, vapes, dip, patches, and gum.

What Type of Drug It Is: Stimulant.

What It Does: At low doses, nicotine can improve fine motor control, attention orienting, and short-term and working memory.

How It Does This: Nicotine acts as an agonist at most nicotinic acetylcholine receptors in the brain—which means that it mimics the binding of our own neurotransmitters and activates the neurons. The downstream effects include modulating the release of a bunch of other kinds of neurotransmitters, like dopamine, serotonin, and acetylcholine.

What the Risks Are: Nicotine is highly addictive. While it doesn't seem to be too dangerous for adults in moderation, there is strong evidence that it causes birth defects, and may lead to developmental problems in adolescents. Plus, there's the fact that most nicotine is ingested through tobacco products, which can cause a variety of serious and frequently fatal health problems, including many forms of cancer. We would personally put nicotine squarely in the "Not Worth It" box, and stick with our coffee instead.

HEALTHY BODY, HEALTHY BRAIN

The funniest thing (we think) about the hype around things like nootropics and brain games is the fact that people are so eager to spend money on products that will improve their cognition and mood and yet so often ignore one of the easiest and most scientifically proven ways to protect your brain: simple physical exercise.

Doctors Hate This One Weird Trick

As annoying as it might be to hear this, the number one way to protect your brain is to engage in regular aerobic exercise—around 30 minutes a day, or about two and a half hours a week. It doesn't have to be anything fancy—just something to get your heart rate up, like jogging, swimming, biking, or dancing. Exercise not only improves your focus and problem-solving for at least a couple of hours afterward but also has long-term beneficial effects, up to and including reducing the risk of dementia.

Wait, You Mean It Works?

It's actually kind of wild how profound the effects of regular exercise can be on the brain, but it sort of makes sense. Aerobic exercise improves your cardiovascular health and boosts blood flow throughout your body, including to your brain. Better circulation of blood means more nutrients getting to your brain. At the same time, exercise has a twofold effect of reducing inflammation, which has recently been implicated in a number of neurological disorders (including dementia!) and lowering levels of stress hormones (which play a big role in a lot of those pesky mood disorders).

We're Going to Pump You Up

Exercise affects the brain at the molecular level, promoting the production of molecules that support brain plasticity. It also helps on a systemic level, increasing gray matter throughout the brain. Maintaining a regular exercise habit improves memory, attentional control, cognitive flexibility, and even information-processing speed, keeping you sharper for longer throughout your life. And even if you don't have a regular exercise habit already, there's still time! Lots of research in older adults, and even in those experiencing mild cognitive impairment, has found that beginning an aerobic exercise habit slows cognitive decline as you age. So, get outside, go for a walk, and get your heart pumping!

PUT YOUR BRAIN ON A DIET

Fad diets come and go, but the brain remains the same. Generally speaking, what doctors recommend as the healthiest diet doesn't vary too much, no matter what label influencers slap on it. And perhaps unsurprisingly, the healthiest diet for your body is generally also the healthiest diet for your brain.

MIND for Your Mind

These days, doctors typically recommend the MIND diet for brain health. *MIND* stands for "Mediterranean-DASH Intervention for Neurodegenerative Delay," which is a combo of a Mediterranean diet and the Dietary Approaches to Stop Hypertension (DASH) diet. Sounds fancy, but at its core, the MIND diet says that you should be eating more of the things you already know you should be eating (leafy green vegetables, fruits and berries, lean proteins, and whole grains) and less of the things you already know you shouldn't be eating so much of (red meats, butter, cheese, refined sugar, fats— basically, everything that actually tastes good).

Brain and Body

Just as with exercise, the MIND diet (like other, similar diets) is good for your overall health and supports a healthy heart, which is good for your brain. These foods provide essential nutrients, including various vitamins and minerals and omega-3 fatty acids, that your body needs to keep itself going. They also provide added benefits that might be important for protecting the brain against cognitive decline, like antioxidants and flavonoids. This doesn't mean you can't enjoy an occasional cheeseburger, but it does mean that in general, focusing on a diet heavy on the veggies and fish and lighter on the butter and beer is probably your best bet for a healthy brain well into old age.

GOING KETO

Keto might seem like the latest fad diet for weight loss and gym bros looking to get swole, but the ketogenic diet was originally developed in the 1920s to control debilitating epilepsy in children. The diet works by swapping out foods high in carbohydrates (like starchy veggies, grains, and sugar) for foods that are high in fat, like nuts and butter. This forces the body to switch from using glucose as its main energy source to instead using ketone bodies. This produces a state called ketosis, which reduces the number of seizures a patient experiences. Even a hundred years later, we're still not quite sure why this diet prevents seizures, but some scientists think it could be that the ketone bodies themselves actually act as anticonvulsants, or that it might lead to changes in the levels of the inhibitory neurotransmitter GABA, which might prevent future seizures from occurring.

OKAY, MR. SMARTY-PANTS

IQ tests supposedly measure a person's cognitive ability and give a score that represents their intelligence and future potential. But when we talk about a person's IQ, or intelligence quotient, we may not be talking about the same thing. Psychologists have developed hundreds of tests, like the Stanford-Binet Intelligence Scales, the Wechsler Adult Intelligence Scale, the Woodcock-Johnson Test, the Cognitive Assessment System—the list goes on. Despite what an IQ test might make you think, a concrete measure of intelligence is nearly impossible, given its abstract nature. What are we even talking about when we say "intelligence"?

Unintelligent Design

There's little agreement on the standard definition of *intelligence*, so every test measures something different, though they often fail to measure broader categories of intelligence like creativity and social intelligence. Likewise, IQ tests don't typically account for factors that can impact a person's IQ, including culture, environment, educational access or background, and even nutrition. And on the darker side, IQ tests have historically been used to justify eugenics movements and discrimination against minority groups and disabled folks. Needless to say, these scores can cause harm if used inappropriately.

A Losing Battle

So you might be asking yourself, "Okay, so, if all these tests are insufficient, what's the future of IQ tests?" Here's a thought: What if we just stopped using them? IQ tests are kind of outdated and horrible for predicting a person's future success, and yet that's how they're used in schools and employment services. The goal has always been to measure general intelligence, yet they all fall short. Specific tests for specific tasks seem fine, but let's just scrap IQ tests.

MORE THAN ONE KIND OF INTELLIGENCE?

Psychologists have tried to come up with alternatives to the general-intelligence model, like Gardner's theory of multiple intelligences, which posits that people have different kinds of intelligences, like musical-rhythmic, visual-spatial, verbal-linguistic, logical-mathematical, bodily-kinesthetic, and so on. This means someone could be very intelligent in certain areas that are not traditionally identified in an IQ test. Unfortunately, while the theory sounds great, there's little empirical evidence to support it. However, others have glommed onto this idea and have proposed new kinds of intelligence that have become wildly popular, like emotional intelligence (aka EI or EQ). Developed in the 1990s, *EQ* describes our ability to perceive, control, and assess emotions that's essential for interacting with others in the world. Is it real? Who knows? Some people are more emotionally attuned, but it's debatable whether EQ is a truly distinct form of intelligence or just a skill or personality trait.

HOW TO *REALLY* KEEP YOUR BRAIN YOUNG

From what we hear, old age isn't so bad as long as you've got your wits about you. But how do you keep your brain feeling spry? Unfortunately, there's no magic pill you can take or app you can download to preserve your brain's youth. However, we've got some suggestions that will actually protect your brain. They may sound a lot like the things your mom or doctor would tell you to do, but maybe they were onto something:

Stimulate Your Brain:
Keep your brain as engaged and active as possible. Try learning a new skill, taking a class, doing new things and challenging your mind!

Move That Body:
Like we've said, exercise is super good for you because it brings more oxygen to your brain. It reverses shrinking of the brain's outer layer and staves off cognitive impairment.

Get Some Sleep:
Look, we had a whole chapter on sleep. Do we really need to tell you it's important? Sleep apnea, in particular, is super tied to early cognitive decline, so sleep!

Maybe Take Aspirin:
Some studies suggest that low-dose aspirin may reduce the risk of dementia, but check with your doctor first.

Lower Your Stress:
Chronic stress and anxiety affect memory and decision-making, and may decrease hormones important in protecting your brain from Alzheimer's disease. Time to change jobs.

Have Close Ties:
Strong social networks have been associated with a lower risk of dementia and longer life expectancy, so call that friend you haven't seen in a while!

STAYING MENTALLY HEALTHY

So far, we've been talking a lot about behaviors you can do to keep your brain in check, but what can you do on the cognitive side? Here are a few suggestions to help you maintain your mental health in tip-top shape. First, practice reframing. It's easy to succumb to negative thoughts, but try countering them by noticing things you feel positive about. You might find you feel more positively about your life overall. Second, forgive yourself for mistakes. Increasing your self-compassion can improve your productivity, focus, and concentration. Third, think grateful thoughts. Gratitude increases your sense of well-being and makes you more resilient to life's stressors. And finally, practice visualizing. Imagining positive things like the best parts of your day or a peaceful beach can decrease your stress and make you feel more relaxed. And if all else fails, see a therapist to explore specialized techniques just for you.

KEEP IN MIND

When you think "meditation," you might think of yoga and juice cleanses, but it isn't just a New Age trend. Meditation has existed for thousands of years and is deeply ingrained in religious and cultural practices from around the world, and there's a reason it's stuck around so long: It turns out that it might actually be pretty great for our brains!

Focus on the Now

Most research has been focused on mindfulness meditation, a technique in which a person focuses their attention on one thing in their current environment, such as the pattern of their breath, or the sound of waves on the beach. This is meant to encourage you to live in the present moment, with a focus on the world around you, instead of dwelling on the past or worrying about the future. Psychologically, mindfulness is associated with improved focus, cognitive flexibility, and greater emotional control, and it appears to be beneficial for mood disorders like depression and anxiety.

A Mind Full of Mindfulness

These changes aren't just mental—they're measurable. When scientists record the brain activity of people who are meditating using electroencephalograms (EEGs), they find that mindfulness leads to increased alpha wave and theta wave activity, which are associated with relaxation and daydreaming. fMRI studies have linked meditation to increased brain activity in the cortex, which processes sensory information and higher-order cognitive processes. These studies haven't been perfect, plagued by small sample sizes and a lack of long-term follow-up, but the evidence so far indicates that a little mindfulness never hurt anybody!

STRETCH YOUR BRAIN LEGS

If you'd like to try mindfulness yourself, here are a few short activities you can use:

The 5-4-3-2-1 Exercise: Take a moment and engage your senses. Notice five things you can see, four things you can feel, three things you can hear, two things you can smell, and one thing you can taste.

Deep Breathing: In a quiet space, close your eyes and take deep breaths. Focus on your breath as it moves in and out of your body. Feel your chest or stomach expanding and contracting as you breathe.

Mindful Eating: Eat slowly, without distraction. Appreciate your food and notice how your body feels while eating. Notice the details of your food that you normally overlook: the colors, smells, sounds, textures, and flavors.

Body Scan: Starting at your feet, scan for discomfort or pain. Acknowledge that sensation and breathe into it. Visualize the tension leaving your body through your breath and evaporating. Move onto the next part of your body when ready.

BELIEF IN A HIGHER POWER

It might seem easy to write off concepts like hypnosis, seances, and past-life regression as made up nonsense peddled by snake oil salespeople to trick folks out of their hard-earned money, but do these phenomena have any basis in reality?

You're Getting Sleepy

While the whole "Bark like a dog, quack like a duck" mind-control aspect of hypnosis is a little suspect, hypnosis does seem to have something of a scientific basis. As we've already discussed, different techniques, like drug use and meditation, can induce unusual brain activity that leads to altered states of consciousness. Similarly, hypnosis is a focused, trancelike state, and it's actually used in some psychological practices. By guiding a patient to focus on a particular object—like, say, a swaying pocket watch—a hypnotist can induce a state of focused relaxation where people become highly suggestible. While under hypnosis, people tend to have increased theta waves, just like those seen in meditation, which are linked to attention and visualization. Other phenomena, like the placebo effect, may be playing a role in why hypnotized people are so suggestible; basically, they're expecting to behave in a certain way while they're hypnotized, so they do. But believe it or not, hypnosis has shown some effectiveness in helping people stop smoking or manage pain.

Are You There, God?

Isn't it kind of odd that as a species, we've repeatedly, independently arrived at the belief that there are higher powers in the universe, or that there's life after death? Well, yet again, our brains might be at least somewhat responsible. Outside of near-death experiences (see chapter 18), there are other clues about the role our brain might play in some religious experiences. Many of the original indigenous cultures that used psychedelic substances, like ayahuasca and psilocybin, used them as key components of their religious ceremonies as a way to commune with the gods, and even casual users today sometimes report profound spiritual experiences. Without the use of substances, individuals with schizophrenia sometimes report religious elements to their hallucinations. There's even a device called the God Helmet that stimulates the user's temporal lobes and, apparently, can induce a "sensed presence," which might take the form of God or the Virgin Mary, or among some agnostics, aliens from outer space. We're not necessarily saying that our brains invented religion to explain weird psychiatric phenomena, but if our entire unique self exists inside our own mind, isn't it possible that our perception of a higher power does, too?

CREDITS

All images courtesy of **Shutterstock**, with the following exceptions:

John Abbott: 109 (Jones-Marlin)
Alamy Stock Photo: 60, 68, 92
Alie & Micah Caldwell: 67, 143 (elephants), 224
Shane Coker: Author Photos
Chris Keeney (Salk Institute) under a CC-By-SA 4.0 license: 107 (Allen)
Getty Images: 64 (Bowlby)
Getty Images/JHU Sheridan Libraries/Gado: 66 (Ainsworth)
Giantnanoassembler (Wikipedia, CC-By-4.0): 108 (Tsao)
Alex Graudins: 255
Elizabeth Loftus: 108 (Loftus)
National Institutes of Health (Public Domain): 109 (Colón-Ramos)
NYTimes.com/Redux Pictures: 106 (Ben Barres)
Oregon Health & Science University: 109 (Fair)
ScienceSource.com: 58 (Erikson)
Scott Eisen (HHMI): 107 (Stevens)
Scott Erwert: 15, 22-23, 27, 35, 39, 46-48, 50, 52-53, 55, 62, 73, 85, 90, 94-95, 103, 121, 127, 152, 154, 157, 161, 170, 173, 177 (GHB), 191, 193, 197, 201, 208-209, 213, 215 (Ketamine), 233, 241, 245, Back Cover illustration
Shane Liddelow: 107 (Liddelow)
Carla Shatz: 108 (Shatz)
Stanford Historical Photo Collection: 99 (prison experiment)
Kay Tye: 109 (Tye)
University of Michigan: 108 (Akil)
Wikimedia Commons: 14, 16-21, 26, 28-32, 34, 36-37, 40, 43-44, 54, 69, 71 (Bandura), 72, 74-82, 86, 88-89 (Watson, rat, baby), 91, 93 (Skinner), 97-99, 107, 110-111, 128, 135, 152, 163 (Berenstain Bears, Sinbad), 175, 221-222, 227, 235, 240 (ECT machine)

Alison (Alie) Caldwell has a B.S. in Brain and Cognitive Sciences from MIT (2011) and a Ph.D. in Neurosciences with a Specialization in Anthropogeny from UC San Diego (2019). Her dissertation research focused on how glial cells known as astrocytes affect the way neurons grow and develop in genetic neurodevelopmental disorders. During graduate school and beyond, Alie has freelanced for other media outlets and has given talks on careers in neuroscience, using social media

for career development in STEM, and the utility of video as a science communication platform. She is now the senior science writer at the University of Chicago Medicine.

Micah Caldwell received his B.A. in Psychology and Spanish from Saint Louis University (2011), an M.S. in Clinical Mental Health Counseling from Marquette University (2013), and is a Licensed Professional Clinical Counselor in California (LPCC) and Illinois (LCPC). As a therapist, he has worked with a diverse array of clients and populations, including individuals recovering from addictions, the homeless in San Diego, recently-incarcerated adults with severe mental illnesses, and unaccompanied children seeking asylum in the United States. He now works remotely as a therapist for college students at Acacia Counseling & Wellness.

About Neuro Transmissions

Everyone has a brain, but unless you've taken an AP Psychology course, not many people have the chance to learn about it in depth. At Neuro Transmissions, we believe that science is for everyone—and we're dedicated to helping you understand your brain! Since September of 2015, we've produced videos about the brain, reaching over 100,000 subscribers. Our work has been the recipient of a number of awards, including the 2016 Khan Academy Talent Search Grand Prize and first place in the 2017 BrainFacts.org Brain Awareness Video Contest.

Alie: I'd like to thank my parents (and sister!) for supporting my love of science over the years, my grandfather for helping me fall in love with science in the first place, and my grandmother for never giving up her belief that I am truly a writer at heart. I'd also like to thank my scientific mentor, Nicola Allen, for teaching me how to be a scientist; and my talented, thoughtful, and generous scicomm colleagues and mentors, especially Mónica Feliú-Mójer, Rose Hendricks, Ashley Juavinett, and Heather Buschman for helping me figure out how to step off the "traditional" STEM path and forge a new career of my own. Most of all, I'd like to thank Micah, whose dedication, patience, and focus is the only reason Neuro Transmissions is what it is today. You're the best partner a girl could ask for.

Micah: I would like to first thank my parents, Patrick and Stephanie, for always cheering me on at every weird hobby I've ever expressed interest in. You've always believed in me. Extra credit to my brothers: David for always asking me the best psychology questions, Ryan for introducing me to moviemaking, and Ian for acting in my first films. I'd like to express deepest gratitude to Dino Arestegui, Ken Gailey, Shelly Falconer, and Soraiya Khamisa for being such important mentors in my career. And finally, my undying adoration and appreciation for my wife, Alie, who has the persistence, smarts, and organizational skills to pull off any dream the two of us have come up with. I love you more than I can actually express in words.

We both owe a huge debt to our editors, Mariah and Ian, for shepherding this project through our cross-country move, our job changes, and a global pandemic. And to our former assistant editor, Madeleine, who is the reason Weldon Owen learned about us in the first place. We'd also like to thank our incredible community of subscribers. It still blows us away that over a hundred thousand of you think that what we have to say is important and that you want to hear more of it. Thank you for watching, and thank you for reading! Until our next transmission—over and out!

Citations

We couldn't put all our citations in the back of this book because that would require a whole bunch more pages and our editors said we've run out of space. For a chapter-by-chapter list of citations, visit https://www.neurotransmissions.science/works-cited/ or use the QR code below.

CEO **Raoul Goff**
VP Publisher, Weldon Owen **Roger Shaw**
Associate Publisher **Mariah Bear**
Editorial Director **Katie Killebrew**
VP Creative **Chrissy Kwasnik**
Art Director **Allister Fein**
Project Editor **Ian Cannon**
Designer **Scott Erwert**
VP Manufacturing **Alix Nicholaeff**

Weldon Owen would also like to thank Madeleine Calvi for the initial inspiration, Marines Alvarez and Mark Nichol for their editorial expertise, and Kevin Broccoli of BIM for the index.

P.O. Box 3088
San Rafael, CA 94912
www.weldonowen.com

weldon**owen**

 Library of Congress Cataloging-in-Publication data is available.

ISBN 978-1-68188-563-6

Printed in China

First printed in 2021

2021 2022 2023 2024 • 10 9 8 7 6 5 4 3 2 1